REPLENISHING THE EARTH

Replenishing the Earth

THE RIGHT LIVELIHOOD AWARDS

1986–1989

Edited by
Tom Woodhouse

GREEN BOOKS

First published in 1990 by
Green Books
Ford House, Hartland
Bideford, Devon EX39 6EE

This edition © Right Livelihood Awards Foundation 1990

All rights reserved

Typographic design and photosetting
in 10 on 13 pt Mergenthaler Sabon
by Five Seasons Press, Hereford

Cover design by Linda Wade

Printed by Hartnoll Ltd
Victoria Square, Bodmin, Cornwall

Printed on recycled paper

British Library Cataloguing in Publication Data
Replenishing the Earth:
the Right Livelihood Awards 1986-1989
1. Social development
I. Woodhouse, Tom
361.6

ISBN 1 870098 38 2

I am grateful to Vince DiGirolamo who provided a fast and accurate typing service, and whose critical comments were valued. I am also grateful to the Barrow and Geraldine S. Cadbury Trust for providing a grant which assisted in the preparation of this book.

Tom Woodhouse

CONTENTS

Introduction ... 9
The Right Livelihood Awards ... 13
Paul Ekins (1989 Introductory Presentation)
The Values of Right Livelihood ... 17
Introductory Addresses of Jakob von Uexkull, 1980-1989

1: TOWARDS A PEACEFUL PLANET

Introduction to the Projects ... 59
Making the Future ... 63
Robert Jungk (1986)
Conscientious Objection and Nuclear Secrecy ... 67
Mordechai Vanunu (1987)
The Global Challenge of Disarmament ... 75
Hans-Peter Dürr (1987)
Peace Studies: Inspiration, Objectives, Achievement ... 89
Johan Galtung (1987)

2: THE RIGHTS OF THE PLANET: GOOD ECOLOGY AND SUSTAINABLE DEVELOPMENT

Introduction to the Projects ... 95
Appropriate Technology and Co-operative Culture in Ladakh ... 99
Helena Norberg-Hodge (1986)
Address to the Award Ceremony ... 110
Tsewang Rigzin (1986)
Chipko—From Saving the Forests to the Reconstruction of Society ... 111
Sunderlal Bahuguna (1987)
Indian Woman and her Role in the Chipko Movement ... 123
Indu Tikekar (1987)
Food First: Beyond Charity, Towards Common Interests ... 127
Frances Moore-Lappé (1987)
The Rainforest of Amazonia and the Global Climate ... 139
José Lutzenberger (1988)
Africa's Food Security: The Preservation of Genetic Diversity in Crop Plants in Ethiopia ... 143
Melaku Worede (1989)

3: THE RIGHTS OF PEOPLE

Introduction to the Projects — 151

The Tribespeople of the Amazon — 155
Evaristo Nugkuag Ikanan/AIDESEP (1986)

The Sarawak Natives' Defence of the Forests — 165
Mohamed Idris and Khor Kok Peng
Friends of the Earth, Malaysia (1988)

The Treatment of Torture Victims — 173
Inge Kemp Genefke/RCT (1988)

Tribal Peoples and Survival International — 183
Stephen Corry/Survival International (1989)

A Statement to all the Peoples of the Earth — 195
Davi Kopenawa Yanomami (1989)

4: PEOPLE'S KNOWLEDGE: THE HEALTH AND DEVELOPMENT OF COMMUNITY

Introduction to the Projects — 203

Nucleogenic Illness — 209
Rosalie Bertell (1986)

The Oxford Survey of Childhood Cancers — 217
Alice Stewart (1986)

Science from the Third World: The Example of Endod in Overcoming Obstacles to a New Approach to Community Health — 229
Aklilu Lemma (1989)

Address by Dr Lemma's Co-recipient — 238
Legesse Wolde-Yohannes (1989)

Building Communities: People's Housing — 241
John Turner (1988)

The Seikatsu Club: Women and Co-operative Community in Japan — 251
Machiko Yajima and Nobuhiko Orito (1989)

Appendix: The Projects and Award Winners 1980-1985 — 261
Bibliography — 283

Introduction

We live in a period of global confusion and doubt. Practical, replicable projects dealing with the challenges facing us are few and far between. This Award is for such projects, the corner-stones of a new world which we can enjoy living in.

Jakob von Uexkull

WAR AND THE ARMS RACE, poverty and unemployment, resource depletion and environmental degradation, human repression and social injustice, inappropriate technologies and potent scientific knowledge untempered by ethics, cultural and spiritual decline: these have often been rehearsed as the crucial problems of contemporary humanity. Their vastness and complexity breed confusion and a sense of impotence.

The Right Livelihood Award was introduced in 1980 to honour and support work which squarely faced these problems and which pioneered solutions to them. Between 1980 and 1989, forty people and projects have received Awards, chosen from 250 nominations from some fifty different countries. Originally funded through an endowment from Jakob von Uexkull, today the Award is funded partly through endowment income and by donations from individuals all over the world. The Right Livelihood Foundation is a registered charity with representatives in England, Sweden, and the United States. It is not associated with any political or religious group. It is open to anyone in any country to make a nomination for the Award, as long as they are not personally a part of the project nominated. The recipients are chosen by an international jury panel. Among its recent and present members are: Walter Bgoya, Director, Tanzania Publishing House; Rodrigo Carazo, ex-President of Costa Rica, and Founder President of the UN Peace University;

Monika Griefahn, Co-founder of Greenpeace Germany; Birgitta Hambraeus, Member of the Swedish Parliament; Sven Hamrell, Director of the Dag Hammarskjöld Foundation; Thor Heyerdahl, author and explorer; Richard Jolly, Deputy Executive Director, UNICEF; Robert Muller, former Assistant-Secretary General of the United Nations; Vithal Rajan, Deccan Development Society, India; Michaela Walsh, President of Women's World Banking; Ponna Wignaraja, former Secretary-General of the Society for International Development; Tom Woodhouse, University of Bradford Department of Peace Studies.

An account of the Awards made between 1980 and 1985 was published by Green Books in *People and Planet* (1987). This volume brings the story up to date and presents the speeches of those seventeen people and projects which won Awards between 1986 and 1989. In addition to the work of the Award winners this volume also presents for the first time the speeches of Jakob von Uexkull, who had the original idea to establish the Right Livelihood Foundation and who, as its Director, has guided it through its first ten years. Born in Uppsala in Sweden, the son of a Baltic-German father and a Swedish mother, he studied PPE at Oxford University. He later became a philatelic expert and dealer, and from this business came the initial funding for the Foundation. In 1984 the German Green Party nominated him as one of their alternating representatives to the European Parliament.

The Right Livelihood Awards are presented annually in the Swedish Parliament in Stockholm. Since 1986 the ceremony in Stockholm has been followed by a seminar in Bradford University, when the Award winners discuss their work in detail. Each presentation ceremony is introduced by von Uexkull, who highlights in his speeches some of the main problems and the values and approaches which the project winners have brought to their work. Collectively these speeches and seminars present a powerful insight into the developing values and philosophies of Right Livelihood, and the reader will find them an illuminating introduction through which to understand both the particular significance of the Award winners and the general orientation of von Uexkull's approach. The

INTRODUCTION

book is organized thematically rather than chronologically and the seventeen different projects are grouped in four distinct chapters: *Towards a Peaceful Planet*; *The Rights of the Planet: Good Ecology and Sustainable Development*; *The Rights of People*; and *People's Knowledge: the Health and Development of Community*. Within each chapter the Award winners (sometimes individuals, sometimes organizations) are introduced with a brief outline of the background to and major features of their work. At the end of the book a profile of those twenty-two organizations which won the prize between 1980 and 1985 is presented.

The collective message of these initiatives is one of hope and reassurance. Today's problems are not insoluble, nor is their solution beyond the resources of individuals and small groups of people acting locally and collectively, mobilizing the energies and talents of others and working for the common good. An important purpose of the Award is to project this message, in addition to supporting the initiatives themselves and disseminating the important knowledge and experience they embody. The Right Livelihood Award aims to stimulate a debate about the values underlying our society and goals. Before we ask 'How to?' we must ask 'What for?'. When 'knowledge' is doubling every two years, we must look at the quality of the knowledge we honour and support. The Award aims to seek out those whose knowledge leads to self-realization, and the realization of values.

Those recognized in this way have rejected the short-term, materialistic goals of much of Western science, technology and society, based on the reckless exploitation of the earth's accumulated resources, and fraught with threatening and destructive consequences. They are pioneering or rediscovering principles and practice rooted in a vision of one humanity, a science of permanence, and an ethic of justice and sustainability which accepts our role as caretakers of the planet. Much of this is part of the perennial wisdom of our species and still lives in many Third World and native communities, from whom the West has much to learn.

A vital feature of the Right Livelihood Award is its holistic approach to the challenges of today. It brings together those who are working for peace and disarmament, for human rights and social justice, for sustainable economic development and environmental regeneration, and for human development, whether through the improvement of health, housing, and education, through cultural and spiritual renewal, or through the addition to the stock of human knowledge or benign technologies. The complexities of our world demands such an integrated approach. Its beneficial results have been proved by many of the Award's recipients and nominees. The task now is to provide the support, inspiration and expertise for these results to be multiplied. The publication of this book marks the tenth anniversary of the Right Livelihood Awards. The following section, which is the text of a speech delivered by the Research Director, Paul Ekins, on the occasion of the 1989 presentations, forms an ideal starting point for this book since it is both a retrospective on what has been achieved and a look into the future.

Tom Woodhouse
Bradford
August 1990

1989 Right Livelihood Awards Introductory Presentation

PAUL EKINS
RLA INTERNATIONAL DIRECTOR

NEXT YEAR, 1990, is the Right Livelihood Award's tenth anniversary year. These 1989 Awards are our tenth presentation. In the light of this anniversary, I have been asked to reflect briefly on the Award's objectives, direction, achievement and future prospects.

A tenth anniversary is a good time to take stock: a long enough period of operation for major trends to be perceptible, not too long for them to have become unalterable. In these ten years, forty Right Livelihood Awards have been bestowed on people, projects and organizations in some thirty countries of the world. How do these Awards add up? Are they simply unconnected recognition of the work of disparate individuals, most of whom did not even know of each other's existence before they unexpectedly found themselves in the Right Livelihood family? Or is there a grand design, of which we in the Award itself may not even have been explicitly aware, an underlying pattern to the work acknowledged by the Awards, an emerging new paradigm that can provide consistent, coherent answers to the huge problems now facing humanity across a whole range of issues?

The honest answer is that we do not know. But we believe that there is just such an emerging pattern, a new articulation of what our Jury member, Ponna Wignaraja, who spoke here last year, calls 'the people-people relation and the people-nature relation'. These new relations will, for example, provide for, indeed will promote cultural diversity within a global perspective, rather than Western industrial hegemony; an ecocentric perception which places humanity within and as part of nature, rather than as external and superior to it; and the development of people in the round, both as individuals and as members of social communities, a development which, in the analysis of previous Award-recipient Manfred Max-Neef, caters for their needs of being, doing and interacting, as well as for the now dominant consumerist need of having.

Over the next year the Award will be seeking to express its 'grand design' more clearly than heretofore. We are planning to bring together our recipients for the first time for a deep collective reflection of their work. They will then meet senior international policymakers in order to influence both their policies and the premisses on which they are formulated. Articles will be written, books published and TV and radio enlisted to get the message out.

It is an ambitious project, but nothing less than high ambition will do. If the Award was founded in 1980 with a sense of urgency, the last ten years have simply served to intensify this urgency. All the major life-threatening trends have worsened through the 1980s. More is spent on weapons now than when the Award was founded; more environment is being destroyed; more people, and more children, are dying or having their lives irrevocably stunted by the deadly combination of hunger, disease and ecological decline.

But there are important changes in the air. Perhaps the breakthrough in East-West relations really will lead to disarmament and a new regime of co-operative common security. Perhaps the freed economic resources really will be used for socially, technologically and culturally appropriate development programmes, devised and controlled by the people affected by them. Perhaps the new environmental awareness reflected in the electoral success of the European Green parties, and in at least the rhetoric of practically all politicians of whatever party, really will cause the industrial nations to tackle the environmental crisis that is so largely of their making: cutting their consumption of environmentally critical resources, reducing their emissions and investing massively in the conservation and regeneration of threatened ecosystems worldwide.

Like all those with our objectives, we will have to work hard to help turn this new awareness into hard policy and direct action. That is part of the challenge of the next ten years.

FURTHER INFORMATION

For further information about the Right Livelihood Foundation contact

Kerstin Bennett
Right Livelihood Award Stiftelsen
P.O Box 15072
S-10465 Stockholm
Sweden

FURTHER READING:

IN ENGLISH
People and Planet, the Award speeches of 1980-1985, edited by Tom Woodhouse (Green Books, 1987); available from the Right Livelihood Office.

IN GERMAN
Der Alternative Nobelpreis, Jakob von Uexkull (Dianus Trikont, 1986, out of print). A new updated edition will be published by Raben Verlag (München) in the autumn of 1990.

IN ITALIAN
Il Premio Nobel Alternativo, Jakob von Uexkull (Edizioni Mediterranee, Roma 1988).

IN SWEDISH
Vi och vår jord, edited by Tom Woodhouse (Energica, 1989); available from the Right Livelihood Office.

The Values of Right Livelihood
The Right Livelihood Awards 1980-1989
Introductory Addresses
JAKOB VON UEXKULL

1980

RIGHT LIVELIHOOD is not just a theoretical moral concept, but a call to us all to begin to take responsibility for the consequences of our actions and activities on our world and surroundings, and to create a society in which this is practically possible.

The outer limits of economic growth are daily becoming more apparent in our polluted and poisoned surroundings. The inner limits of purely material growth show themselves simultaneously in the increasing unhappiness with the spiritual poverty in our rich world where every second hospital bed is already occupied by a mentally ill person.

A few years ago E. F. Schumacher called on us to create a society on a human scale, instead of a society adapted to the needs of machines. 'Technology is on its way to reach such a perfection that human beings believe they can do without themselves,' writes Polish author Boleslaw Lesz.

Today it is obvious that the consequences of continued economic growth will be negative and catastrophic. Serious problems will become insoluble. If we wait till the year 2000 before we accept new values and ways of life, it will be too late. According to the Club of Rome Report, the equilibrium between nature and humanity will by then be irretrievably lost, even if this may not yet be apparent.

A healthy organism grows until it has reached its optimum size, after which it matures and steadies. Unlimited material growth is the sign of a sick organism. If our ancestors had lived as wastefully as we do, we would not be here. If they had patented their knowledge in order to sell it dearly (as we do today to the Third World), we would be poor today.

We talk about a knowledge explosion today but what we have is a knowledge inflation: an enormous quantitative growth of qualitatively even poorer pieces of knowledge, while valuable traditional knowledge is often ignored and forgotten simply because it is 'old'. Erwin Chargaff foresees us progressing to a state in which we will have so much lead in our brains that no one can understand the computer programmes which control our lives. We have come so far in our admiration for the eternally new that we do not even have words for the opposite sentiment. Chargaff tries: 'If I were younger', he writes, 'I would found the Coverers' and Disinventors' Club.'

The Right Livelihood Foundation wants to support integrative, holistic, healing knowledge as opposed to divisive, mechanistic, reductionist knowledge. Today's scientific system, says the former president of the West German Association of Scientists, Klaus Meyer-Abich, is the institutional expression of a growing deficiency in the perceptive abilities of individuals and of society. Nature is perceived as incompletely as is the human being, seen only as a thinking thing.

The major conflict of our time between the competing materialistic growth systems of East and West is a race to see who can first turn the earth into a moon landscape. Those who in their daily lives are surrounded by the wretched of the earth have little to say in favour of such 'alternatives'. Thus Mother Teresa of Calcutta summarized her impressions after her visit to Europe: 'Here I find a different kind of poverty, a poverty of loneliness, of being unwanted, a spiritual poverty—and that is the worst illness in the world today.' . . .

No technological solutions can supplant the natural equilibrium of our planet, for human technology, like all other human activities, takes place within this balance. Right Livelihood is to 'live lightly' on the earth entrusted to us, not to use more than our fair share of its resources. Infinite growth belongs to the realms of the spirit. Luxury is a fitting attribute not for consumer goods but only when we build for eternity, as our ancestors clearly knew—see their cathedrals and temples!

We must create viable new realities in all areas, alternatives which can show ways out of the chaos ensuing in the not-too-distant future when the techno-bureaucracy comes to a standstill because all its energies are taken up by internal transactions, and the materialist paradise loses the remains of its charm. The Right Livelihood Award is a part of this alternative mosaic, to which in fact there is no alternative if we are to survive.

We want to reward those whose knowledge is healing rather than splitting the world, those who preserve and rediscover the understanding that we are not machines in a dead universe but living participants in an indivisible organic whole. This Award is not for beautiful theories; it is for those who have turned knowledge into responsible practical work in their daily lives.

In order to change our world we only need to change our success criteria. 'Eternal economic laws' become irrelevant statistics when wisdom and spiritual awareness are our most sought-after goods, and when our possessions have ceased to possess us. As soon as the maximization of production and consumption ceases to be our first priority, and is replaced by the desire for a natural balance, for justice, sustainability, self-reliance, co-operation and a whole and integrated personality, then the 'immutable laws' of the materialists have as much relevance as the dogmatic debates of medieval churchmen.

Today more and more men and women in the rich world are willing to make material sacrifices in order to share with the poor and preserve natural resources. Theirs is not the joyless poverty of ascetic fanatics but, on the contrary, the joyous frugality of those who have outgrown the tyranny of infinite wants, who are no longer possessed by their possessions. In a recent study of the priorities of Swedish schoolchildren for their future working lives, 'a high income' came almost at the bottom of the list (ninetieth out of ninety-nine possible choices). At the same time the awareness is growing in the Third World that the Western 'development' path is impossible to follow without reaching its outer and inner limits long before arriving at its apparent desired objectives (see, for example, *Poverty: Wealth of Mankind* by Albert Tevoedjre).

Our increased technological, manipulative knowledge has made us blind to the loss of that creativity and awareness which gave past generations the knowledge of being part of a larger whole. Hassan Fathy and Stephen Gaskin represent the living re-emergence of this awareness and this creativity.

Professor Fathy, architect, inventor, musician and author, has worked for half a century to preserve, adapt and modernize ancient knowledge and skills in the building of houses, using traditional local materials. Long before others, he saw the danger in making our basic need for shelter subject to the rules and whims of the modern market society.

Stephen Gaskin founded the Farm Community ten years ago as a means of preserving and putting into daily practice the idealism of the hippie movement: non-violence, hospitality, sharing. The Farm is an example of that spiritual revolution today which—far from implying a withdrawal from worldly activity—works to save and change the planet by showing in practice that love and sharing, not greed and envy, are the characteristics most in tune with our inner selves.

We see the Awards today as a necessary alternative and complement to the Nobel Prizes. This must not be understood in an antagonistic sense but as an impulse and inspiration. Sir Brian Pippard, a Nobel Prize nominator, complained recently that the scientific Nobel Prizes put far too much emphasis on elementary particles: 'What has the discovery of all these elementary particles to do with the real problems in the real world?' The work of Hassan Fathy and Stephen Gaskin has everything to do with the real problems in the real world today.

1981

This prize is intended to support and honour those whose knowledge is healing, rather than splitting the world, those who know that we are not random mechanisms in a deaf, frozen, indifferent universe, but living participants in an indivisible organic whole. Adequate, meaningful knowledge cannot be understood without being lived.

Our challenge today is less a lack of answers than the failure to practise them. Most of our activities still go on as if the great debates of the past decade had not yet happened. Often it is not the lack of funds but the massive powers of the status quo which are used to stifle new initiatives. The men of power cannot admit their powerlessness, except when a private event jolts them out of their cosy institutions. Klaus Traube, one of the top managers of European nuclear industry until recently, is now one of its most bitter critics. When asked to explain how he could have said the things he did a few years ago, he replies, 'I really cannot explain that myself.'

It is vital to realize that the emperor really has lost his clothes, that the men of power are powerless, that the alternatives are often the only alternative, and not just some fringe activity. We must stop staring at the present. The walls of Jericho looked just as mighty forty seconds before they collapsed as they had done forty days previously! . . .

We all must decide whether we want to be part of the problem or part of the solution. We cannot leave it to the experts. Every one of us has his or her unique contribution to make. As Buckminster Fuller has said, the scales are now so evenly balanced between disaster and survival that every single individual can make a crucial difference.

This year we acknowledge the work of three individuals who have recognized this and acted accordingly. In their chosen fields they have created new realities, working on our three most important challenges: the waste of scarce resources in the arms race, the threat to the ecosystem, and an educational system incapable of fulfilling the aspirations of the great majority. Together they represent science in the service of humanity and life as opposed to the present cult of knowledge, for the sake of knowledge, serving an outdated mechanistic worldview. . . .

In the 1960s our scientific priesthood promised us paradise on earth for the 1980s. Today scientific progress reports are more modest. As Mike Cooley says, it is not surprising that most people today believe technologists to be people who deliberately spray rust on the cars before the paint!

At the same time, our whole world is being disordered and disrupted in the name of an economic efficiency creating short-term profits for a small minority at the expense of us all. It is hard not to agree with the American historian William Irwin Thompson that

'The fact that post-industrial civilization awards Nobel Prizes to economists speaks volumes on just how much we have lost all sense of perspective in even our most ceremonial occasions.'

It should be noted that the economics prize is the only new prize accepted by the Nobel Foundation during the whole history of the Nobel Prizes. The scientific basis for economics can be best summarized by a recent indiscreet comment by the Budget Director of the most powerful economy in the world, Mr David Stockman in the White House. I quote: 'None of us really understands what's going on with all these numbers.'

1982

... Our ancestors saw themselves as embedded in the flow of life, as a link between the past and the future. Three generations inhabited their world: the ancestors, the living, and the yet-to-be-born, whose interests were weighted when decisions were taken. Impossible in such a society to make life more precious for the present at the expense of the future; impossible to develop nuclear power for the present generation and leave the lethal waste to the care of our descendants: impossible to exploit and ruin our soil and heritage. (According to the accounting practices of agribusiness, soil in forty years' time will have no financial value.) ...

The recent works of James Lovelock, Kit Pedler and Peter Russell strongly suggest that our planet is a living being. It is not at the mercy of that random evolution so favoured by small-minded reductionists, who, in the words of E. F. Schumacher, are sceptical of all those values which demand anything from us, but never sceptical of that scepticism which demands nothing at all from us. Our planet has retained a stable, life-giving environment against all odds over many millions of years. It has self-healing powers, like any living being, but our generation has in a few short years depleted a considerable part of its accumulated wealth and health.

The fraudulent promise that one day everyone will live like Swedes or North Americans has been used to justify the exploitation of the natural resources of the Third World. Today we know very well that the modern scientific magic has failed, that more of the same kind of science will not solve the problems it has created, that if everybody began living, consuming and wasting tomorrow like the average Swede does today, we would run out of all our basic raw materials within months rather than years. . . .

Our primary purpose is to help support and spread a more comprehensive, holistic knowledge than that recognized by orthodox science, to recognize a science of being, integrated in our daily lives and actions, which can help us to solve the challenges we face today. Before we can expect useful answers we must learn to ask the right questions.

Traditional sciences, for example, are based on a more comprehensive sense of reality than most of modern science. This does not mean that we should return to bygone ages, which is impossible. But it does mean that, while forging the new, we are aware of our roots and do not believe that everything worthwhile and beneficial was discovered in this century.

We are told that we must continue our eternal competitive rush, that we cannot turn back or even stop the clock. Few of us would buy a clock which cannot be reset when and as required, so I wonder how long we are prepared to put up with a system which has become too powerless, confused, sterile and imprisoned by its own institutions to be able to offer us more than different varieties of a clock gone out of control. . . .

The wisdom of ancient civilizations is being forgotten at the very moment when we most need it, when many young in the Third World queue up in a futile attempt to get on the lowest step of the ladder which we have told them leads to paradise on earth Hollywood-style, but which in fact leads nowhere.

But, despite our onslaught, islands of sanity have survived. Erik Dammann was inspired to initiate the *Future in Our Hands* network after living in Samoa. He found there people as eager to be respected and esteemed by others as anybody else. But in Samoan society you cannot gain respect by actions which give personal power or wealth. Samoans admire only those who do most for others. Similar value systems were prevalent in most human cultures. Only our culture has turned egoism into a virtue, with the predictable results.

The Right Livelihood Award jury felt the time was ripe this year to present an Honorary Award, for this Award is not just monetary support but an expression of recognition, of solidarity and of love. I am very happy that Erik Dammann and the *Future in Our Hands* network is the first recipient of this Honorary Award. I think it is obvious that there can be no change in the world if we in the materially rich nations do not change our lifestyles. The *Future in Our Hands* is the first and most successful such movement in Norway. It is represented here by Leif Sandholt, who is also the initiator of an organization bringing together educators and teachers to further the teaching of ecological responsibility and global solidarity in Norwegian schools and has founded a project involving direct co-operation between localities in Norway and Third World countries....

I am often asked what caused me to take this initiative. I can find no better way of expressing my motivation than to quote Erik Dammann's answer to the same question. It was not really unselfish idealism, he says. 'It was more a question of our own needs. We had to contribute to the creation of a world which we ourselves could stand living in.'

The movement towards a more complete vision of reality and human nature is the fastest growing development in our world today. At the crest of this wave stands Sir George Trevelyan and the Wrekin Trust, one of the major influences in the revolution in thinking which is beginning to enlighten the world. Countless lives have been changed for the better by the work of the Wrekin Trust. It is not tied to any dogmas or religions. We do not need any new religions today, but we do need religion in the original sense of the word—religare: to reconnect ourselves with our source and with our world, to understand that the needs of the person and the needs of the planet are today the same, and to live accordingly.

How many people have entered the hectic and exhausting political arena because they see themselves as representing not just their voters, not only the minorities, the ill and the socially weak—but the children, the coming generations, the animals and the plants as well? This is the task which Petra Kelly has taken upon herself. Extraordinary dangers demand extraordinary efforts and I know no one who devotes more effort to open up the sterile fossils of the institutions controlling our lives. Hiroshima is still in our minds as a day of unsurpassed human infamy. Today's nuclear weapons are enough for one Hiroshima every day for the next 4,000 years! . . .

Petra Kelly said recently: 'It is time to become and to be both tender and subversive.' The Award to her is as much for her tenderness as for her subversiveness. . . .

Such endeavours of the movements in the industrialized world are paralleled in the Third World by the work of which PIDA is in the forefront: learning how to generate self-participatory self-reliance, neither tranquillizing the oppressed, nor just mobilizing anger, but inspiring and empowering the poor, helping them to ask the questions to get the information which will enable the process of change to begin; and all the while swimming (as the Chinese say) not against but between the currents.

While PIDA aims to build self-reliance, the work of Anwar Fazal is aimed at protecting those who have been tempted away into the global supermarket, subjected to a trade which, in Anwar Fazal's words, is more like an armed attack. People in Third World countries, he says, 'had a vision that big company executives were like priests, that these people had a certain morality, a certain responsibility, because they came from higher levels of civilization and from more powerful strata of society.' Indeed, in other civilizations power and responsibility went hand in hand. In medieval Japan the emperor dispossessed the rich if they lived 'so as to cause envy among the people'.

Through Consumer Interpol, a global investigative network to report on and prevent the export of hazardous products, wastes and production processes, Anwar Fazal works to create responsible consumers and producers, aware of the impact of their actions.

Our Award recipients this year come from very different backgrounds and fields of expertise. What unites them is living and working to make whole instead of splitting our earth, to heal instead of destroying, to uplift and fulfil instead of reducing life.

1983

. . . We aim to reward work which is ecologically responsible and which does not ignore the traditional wisdom of mankind, especially the knowledge of the Third World. For a truly international award has to break out of our cultural straitjacket, our prejudice which assumes that everything worth knowing and doing has been discovered in Western universities in the past fifty years.

We are in a period of global confusion and doubt which has gripped even our most stable institutions. New theories abound but practical, replicable projects dealing with the challenges facing us are few and far between. This Award is for such projects, the corner-stones of a new world we can enjoy living in.

The other purpose of this Award is to stimulate a debate about the values underlying our daily work, our goals and our institutions. We must identify, expose and question these values. Science and technology promise us answers to the 'How to?' questions but we no longer have an adequate answer to the 'What for?' question. Without such an answer the wonders of modern science will perish as did the highways and aqueducts of the Romans.

The 'crisis of meaning' paralyses our efforts to deal with the other crises facing us. The West German politician Erhard Eppler writes: 'The conventional wisdom of our society claims to be objective, value-free, following only the dictates of reason and fact. The sooner and the more often we expose this dangerous nonsense the better.' To quote Gunnar Myrdal: 'Values are always with us. Disinterested research there never has been and can never be. Prior to answers there must be questions. There can be no view except from a view-point. In the questions raised the viewpoint has been chosen and the valuations implied.' Or, as Albert Einstein said: 'It is the theory which decides what we can observe.'

So we are not trying to introduce our subjective values into a world of objective realities. Rather, we have taken a long hard look at the hidden value assumptions behind these supposedly objective realities and we reject them as unworkable and abhorrent. . . .

The sociologist Max Weber foresaw an 'icy polar night of society where experts without a soul ruled hedonists without a heart.' While twenty years ago fifty-eight per cent of Canadian parents expected their children to have a better life than themselves, today these figures are reversed: eighty-five per cent expect their children to have it worse. In a recent Swedish study, eighty per cent of the young people questioned felt pessimistic and powerless. An eight-year-old girl was quoted as wishing that she could be old like her grandfather so that she did not have to worry so much about the bomb any more. A science which proudly claims to be the 'carrier of all promises on earth', but refuses to take responsibility for having shaped a world in which children wish that their life was already over is an affront to our humanity. The biochemist Erwin Chargaff writes: 'Those who today face science, without having to live off her, can hardly escape the impression that they are being killed by her.'

And if the victims of this science turn in their despair to the medical sciences for help, they are likely to find, as did the American writer Joanna Macy, that her 'deep dismay over the destruction of the wilderness was diagnosed as fear of [her] own libido, which the bulldozers were taken to symbolize, [and her] painful pre-occupation with US bombings in Vietnam ... interpreted as an unwholesome hangover of puritan guilt.'

The ultimate consequence of modern science looks increasingly likely to be what scientists claim to be the goal of their critics: a return to stone-age conditions, not, of course, through the actions of environmental fanatics, but through the very opposite, namely our inability to put enough sand in the machinery of destruction.

After the Seveso disaster in Italy, the anger of the victims turned not only against the company responsible but against all authorities and outsiders, especially the scientists who came to 'help' but were soon forced to admit their ignorance. After the US bombings of Cambodia, the survivors turned not just against the servants of the fallen regime. Wearing glasses was sufficient reason to be regarded as an associate of the modern technology which had destroyed their country, and to be killed by the Khmer-Rouge. If our scientists expect to survive a nuclear catastrophe in their bunkers, how do they expect to escape the fury of the people when they come out? Perhaps such reminders are needed to increase the so far tiny trickle of courageous scientists who are refusing to serve the forces of destruction any longer.

The critics of science are accused of Luddism. The Luddites, we are told, went around England during the early days of the industrial revolution, destroying machines, because of their blind hatred of progress. In fact they only destroyed those machines whose owners had reduced the wages of their workers and/or the quality of their produce. Thus, far from being blind, the Luddites acted consciously against the deterioration in the quality of life in the name of a new order benefiting only a few..

The blind destroyers today are the fanatics of progress who ignore, destroy and disdain the fruits of the accumulated wisdom of mankind. Hassan Fathy, our first Award winner, has worked all his life to preserve the traditional architecture of his native North Africa where mud-brick buildings have been continually inhabited for up to 4,500 years. Of course, he said, they can be improved with the help of modern knowledge. But instead, modern architecture rejects them as 'primitive' and introduces Western technology, which for the Third World often means at best a concrete box with a corrugated iron roof under which children die of heat exposure.

One of the very few scientists from the Third World to receive a Nobel Prize is Abdus Salam. In a recent interview he declared that, in order to achieve real independence, Third World countries must build, despite enormous cost, their own nuclear accelerators to conduct their own research in the prestigious world of sub-nuclear physics. Within the present system, he is, of course, right. Only when Third World scientists start discovering elementary particles in their own accelerators will they be taken seriously by our scientific establishment—if it still exists. . . .

Instead of our short-term science and technology, based on the quick exploitation of the earth's accumulated non-renewable resources which leaves us powerless when faced with the consequences, we need a science of permanence, a science based on the perennial wisdom of mankind, a science which accepts our role as caretakers of the planet. We need a search for knowledge leading to self-realization and the realization of values. . . .

What is needed today is 'not one more variable, or dimension, or even perspective', says Rajni Kothari, 'but rather a new comprehension of the human *problematique* which entails the evolving of an alternative paradigm of scientific endeavour.'

The work of Professor Leopold Kohr plays a vital part in this new paradigm. Long before anyone else he questioned our belief that if something is good, more of it must be better. . . .

Over twenty years ago Professor Kohr published a book entitled *The Overdeveloped Nations*. He showed that it is not our problems which are unprecedented but their size: a return to a more human scale in our institutions and communities will make our problems manageable again. . . .

To those of us who have grown up and lived in Northern Europe the islands of the Pacific have always held a special fascination. Especially in winter we look at the opposite side of the globe and dream of the South Seas. Whenever I met someone who had sailed round the world I would ask what was the most beautiful place they had ever been to. If they had been to Palau, the answer was always—Palau.

Today the name Palau stands not just for beauty but for courage, inspiration and perseverance. To those who say that individuals and small nations are powerless, we say: look at Palau! Rarely has there been a more unequal struggle. One of the smallest nations in the world has asked for no more than to remain free of nuclear arms, waste and military bases. The most powerful nation on earth has repeatedly betrayed the democratic ideals it claims to uphold, attempting by all the means and tricks at its disposal to force Palau to abandon its status as the world's first constitutionally nuclear-free nation. A more blatant disregard for the welfare of a people who have been temporarily entrusted to the United States by the international community is difficult to imagine!

We are told by our experts that we need yet more economic growth, that the rich must get even richer so that they can buy from the poor and afford to help them. The growth rates declared necessary to combat our unemployment mean that the per capita income in the rich countries must increase every year by more than the total per capita income in many poor countries. At this rate, every Swede would have his or her own private jet before the poor have provided for their basic needs. Has there ever been a human society based on a belief system so absurdly ridiculous?

Professor Manfred Max-Neef writes: 'It is no longer a question of correcting what already exists. That opportunity was lost long ago. It is no longer a question of adding new variables to the old mechanistic models. It is a question of remaking many things from scratch and of conceiving radically different possibilities.'

The figure of six billion dollars used to signify two things. It was the sum needed, according to the US President's Commission, to end world hunger. It was also approximately the US military budget for one week. Recently, it has come to signify something else: it is the estimated cost of a nuclear power station in Washington State which was abandoned before it produced any electricity.

The enormous cost of nuclear power as well as the ecological and social dangers, and the advantages of renewable energy and conservation have often been pointed out, but never more forcefully and persistently than in the work of Amory and Hunter Lovins. Their figures have often been challenged but never disproved. They have been consulted by governments and by companies and investors seeking a way out of the quagmire into which their addiction to technological fixes has led them. But Amory and Hunter have first and foremost put their vast knowledge at the service of the people, helped to translate and expose the experts' jargon and made the energy debate accessible to its main victims, the consumers. . . .

I would like to conclude with these words of Erhard Eppler, one of the few politicians in Europe today whose horizon is not the next election but the next generation: 'There exists no wide road into the future except into catastrophe. A blue-print for the world in 2010 does not exist. There exist only tracks, and for the moment they are leading through thorny bushes. Now one begins to work one's way in a certain direction, tearing one's trousers and making one's fingers bloody. But there is a thread and needle for repairing, adhesive plaster to put round the finger. For my part, I have used a lot of adhesive plaster, and it does no harm at all. One may improve a track, and the more it is used the sooner it will become a path.'

1984

I would like to thank all those, especially the Swedish Right Livelihood Society, who arrange these presentations and manage to find the golden mean between organizational perfection and the openness and spontaneity this Award stands for. . . .

The futurist Arthur C. Clarke writes in *Profiles of the Future* that our knowledge is doubling every two years. Already, he says, twenty years' schooling is insufficient, and soon we will die of old age before we have learnt to live, and our culture will collapse due to its incomprehensible complexity. But knowledge is not quantitative. What valuable knowledge in the Third World is today being lost, suppressed by the mass-produced knowledge which Professor Clarke admits leads to collapse? It was Mark Twain who wrote: 'When they had lost sight of their goal, they redoubled their efforts.'

Western science claims that all spiritual and mental experience is a meaningless by-product of chemical reactions and that we owe our existence to chance mutations. Repeatedly scientists proclaim the discovery of the missing link between humanity and her supposed animal ancestors, only to have to admit some years later that the fossils in question were, if not a fraud like the Piltdown man, then perhaps a monkey but not a 'link'. That we have been created by chance mutations is as probable as the belief that a tornado whirling through a scrapheap long enough would assemble an aeroplane.

Our ancestors had a science of the soul. They knew, perhaps, fewer facts but they knew the crucial fact that they were 'known', that they were at home and not lonely strangers on this planet. This knowledge is contained in their myths. A science which denies reality to this deeper knowledge and seeks to brainwash us, through the media and educational system to accept that their mechanistic explanation is the only possible hypothesis, is in danger of losing all credibility. . . .

Our experts plan a future which few of us want: highly materialistic, industrialized and centralized. A recent survey of British schoolchildren confirms that the majority of the young in the West today want a more just, less materialistic, more decentralized and rural society with fewer chemical additives, and that they are ready to accept simpler lifestyles. This is not a return to the past (impossible anyway, as we have burnt the bridges behind us), but a future very different from that which we are building for them.

When President Reagan appointed Mrs Burford to head the Environmental Protection Agency, she dismissed many of the agency's scientific advisers, declaring: 'they are good scientists but we want our scientists and not their scientists.' It is good to remember those words next time 'their' scientists try to present their findings as objective reality.

Our 1984 Awards go to the women of the Third World, whose problems, lives, and values have long been ignored. Their contributions to human well-being have been ignored by economists, their needs by technologists, and their health by the medical sciences because they do not command sufficient monetary resources. It is a great pleasure to present our Awards to four women whose tireless, inspiring work has moved us all.

Through SEWA, represented by its founder Ela Bhatt, thousands of Indian women today have their own resources, greater freedom, security, and self-respect.

Wangari Maathai has inspired and organized the planting of over 700 Green Belts which have begun to reverse the trend of environmental destruction and improve the soil, the climate, and the human environment in Kenya.

Giving free legal assistance, helping the poor to organize and to claim their rights, enabling the victims of injustice to overcome their bitterness while protecting them from abuse and bringing their persecutors to justice: this is the work of the Free LAVA network founded by Winefreda Geonzon.

The non-sectarian Beirut peace movement was inspired by a poem which crystallized the longings and despair of the hitherto silent majority, the victims of the war both young and old. For her poem, her work, and her courage the Honorary Award is presented to Iman Khalifeh of Lebanon.

I would like to close with the words of a man whose energy and example is now sorely missed, and who showed us that we must change our ways if the planet is to survive. The late Aurelio Peccei, founder of the Club of Rome, said: 'It is within our power to make a better world, a peaceful society. We can do it. So there will be absolutely no excuse if we do not.'

1985

The common themes of this year's Awards are human rights and people's empowerment. The Awards aim to empower by supporting individuals and projects who did not wait for the experts' and bankers' approval before they began to work, but who know that the power to change is in us. Nothing is more inexcusable than to fail to act because we believe we can only do a little. Pessimism and optimism are both disempowering: the one believes action is useless anyway, the other that things will turn out all right, whether we act or not. On the contrary, how they turn out depends on each one of us—we do not know beforehand which snowflake will finally cause the tree branch to crack, which song will cause the walls of Jericho to crumble.

This Award is for those who know their song, who know what they care about and live a life that demonstrates their concern. It is an Award for realists, for nothing can be more unrealistic than to live so-called normal lives in these abnormal times. . . .

In the last few years, through the work of men and women, such as our recipients tonight, there are rays of hope. We have not yet turned the corner or even stopped the drift into global barbarism, but we are becoming aware that there is a corner to turn, that there are other ways. Most of our fellow humans still need to wake up, but we need to ensure that the reality they wake up to is not devoid of hope. That is why it is vital that work such as this becomes known, understood and recognized.

Our diversity is our greatest wealth. Unlike the prophets of modern science and technology, we do not pretend to have the cure for all ills, we do not believe that all our solutions should be globally replicated. But as examples of pathways out of powerlessness and confusion they are invaluable. . . .

It is time to choose. Science can only present us with a view of the world inferior to us, because the scientific method is based on controlled experiments, and what is higher cannot be controlled. Thus, there are no 'objective' criteria, no experts who can make our choices for us. Our choices about which road to pursue can only be based on our values and beliefs. Our social values depend on whether we believe consciousness to be localized and brain-bound or essentially universal, whether we want to conquer and subdue nature 'out there' or whether we want to overcome the limitations of our own nature.

Overcoming these limitations means first of all expanding our awareness to see the injustices perpetuated in our name; it means undoing the conditioning which makes us put up with the criminal obscenities of the destruction and waste around us.

Some believe that spacemen will come and save us, some believe that only by going out into space can we be saved. Neither group wants to face the consequences of their actions. We, on the other hand, believe that our work is here and our time is now. In the words of the British historian Arnold Toynbee: 'During the disintegration of a civilization, two separate plays with different plots are being played side by side. While an unchanging dominant minority is perpetually rehearsing its own defeat, fresh challenges are evoking fresh creative responses from newly recruited minorities who proclaim their own creative power by rising each time to the occasion.'

Tonight we welcome a man who, on numerous occasions, at risk to his career and safety, has risen to the occasion. Professor Theo van Boven works to remind us that human rights are not just a matter of words, that abuses are not statistics but involve a human being subjected not just to a denial of rights most of us take for granted, but to pain, degradation and suffering on a scale which surpasses our comprehension. For bringing the cries of the tortured into the halls of diplomacy, Professor Theo van Boven truly deserves our Honorary Award.

The struggle to preserve our environment, to ensure that development does not remain a password for destruction is now global. For a while we were told that countries which did not know the unfettered pursuit of private profit would find it easier to preserve a healthy environment. Recent news from Eastern Europe shows that this is unfortunately not so. The pursuit of industrial expansion and economic growth at any cost is leading to a decreased quality of life in East and West. Quantitative standards may increase but a closer inspection reveals that the only thing still growing is the social cost of growth. In East and West, groups of concerned citizens and scientists join together to bring into awareness the true costs of the prevailing belief that large-scale technological solutions are automatically beneficial. . . .

Janos Vargha and his colleagues in the Duna Kor network have worked under very difficult circumstances to preserve a vital part of Hungary's natural environment. . . .

Much vital work is being done in isolation by small groups who find it difficult to link up, to obtain the information and protection which would make their work more effective. This is especially the case in the Third World and in a large country, such as India. Therefore, the work of Lokayan, as a network connecting local and regional groups working to protect human rights against abuse, discrimination and pollution, stands out as a vital and creative initiative if we are to shift the inertia of the status quo and the forces of greed, fear and oppression in the world.

Some violations of human rights are clear and immediate, especially to those whose rights are violated. In other cases, even the awareness of what is being lost may be missing until it is too late, but the longer-term consequences can be catastrophic. The dangers of losing our plant genetic diversity, of becoming dependent on a handful of herbicide-laced fertilizer-hungry, badly adapted varieties, controlled by a handful of multinational seed companies, have been brought to the world's attention by two men: Pat Mooney and Cary Fowler. They have not stopped there, but work to reverse this development which threatens a vital part of our heritage.

1986

These Awards are for those who work to ensure the survival of the human and other species on this planet. At present the rate of extinction is about one species (plant or animal) per hour! Every day 50,000 human beings die unnecessarily from hunger and malnutrition. The rich countries today export the consequences of the limits of growth to the poor. We buy our food from the Third World when we have polluted our own. We have acquired the knowledge to annul the future. For the first time in history, the apocalypse can be triggered by humans.

The defenders of the present order with its large-scale, increasingly uncontrollable technologies speak of the remaining risk with which we must learn to live. But many of us feel that the remaining risk of doing without these technologies is more acceptable. Are we timidly to become the victims of our own creation? Are we really prepared to poison ourselves, our children and our earth, and abdicate our freedom, for the comforts of the plutonium society? Who selects the information on which we base our choices? . . .

When the first nuclear bomb was ready in 1945, the US Minister of War asked four leading physicists among its creators (Robert Oppenheimer, and the Nobel Prize Winners Fermi, Lawrence and Compton) to recommend how to use it: should there be a technical demonstration explosion on an uninhabited island, or should it be used militarily straight away? It was the scientists who recommended the 'immediate military application', i.e. the bombing of Hiroshima and Nagasaki. From the available documents it is clear how uncomfortable they felt about being asked. They had created the bomb but did not want to deal with the consequences. Such science we could not and cannot afford.

How could it come to this? How has the pursuit of human knowledge become a danger to human existence? In the mechanistic worldview of our scientific materialists life and nature are nothing but a heap of molecules without meaning or purpose.

The poet Kathleen Raine writes: 'To deny the life, the meaning, the values, in the last instance the holiness of nature is to deny ourselves wings and eyes to experience that world so rich in qualities which respond to those meanings and values which we are imaginatively equipped to comprehend and express.'

Tonight we honour a man who, in the twenty-five years I have known him, and long before, has worked to expose the cover-ups, the fallacies and the arrogance of power which threaten our survival, a man whose warnings and righteous anger have always been accompanied by the conviction that people want to and can take back their future from the experts who have monopolized it. In these difficult times we are indeed fortunate to have Professor Robert Jungk on our side.

This Award is not just given in areas not covered by other awards, but also for pioneering scientific work which goes unrecognized elsewhere, because the findings do not suit the present scientific and political establishment. Even to admit that Dr Alice Stewart could be right would mean not just the immediate closing down of our nuclear industry, but would also amount to an admission by too many that they have been too wrong for too long. Dr Stewart's early discoveries of the dangers of X-raying pregnant women have led to regulations which in the

THE VALUES OF RIGHT LIVELIHOOD

past decades have saved the lives of tens of thousands of young children.

A scientist whose invention had such a beneficial effect should receive the highest international acclaim. But Dr Stewart did not make a new invention from which industry could profit. On the contrary, she has shown that widely-used nuclear technologies are highly dangerous. For the high priests of our age such work is sinful heresy against progress.

The Ladakh Ecological Development Group has shown that the refusal to follow in our Western footsteps does not mean stagnation, that other cultures can choose which technologies to introduce and integrate without destroying themselves, can choose either subservience to high-tech and the world market or a partnership in which knowledge is shared. Living in harmony with the earth, living simply so that others may simply live—these are some of the lessons for us from Ladakh. Dame Barbara Ward wrote of 'the variety and gaiety ... of earlier [European] societies, in which few people worked more than half the year, and the record of songs, dances and tales brings up echoes of a merrier life.' But her medieval Europe must have been a stressful place compared to Ladakh where, I understand, you only work four months of the year!

The modern war against nature is also destroying the skills and wealth of the native peoples. Our definition of skills ignores the most valuable: how to survive without destroying what sustains survival. Our Award-recipient in 1983, Professor Manfred Max-Neef from Chile, spoke of the modern poverty of love, of protection, of understanding, of participation and identity, for which money is a very unfulfilling substitute.

With us tonight are representatives of the native peoples of Amazonia, representatives of cultures rich in those aspects of life to which Manfred Max-Neef referred, but under threat from a Western culture which counts wealth only in money. Evaristo Nugkuag Ikanan is President of AIDESEP, which in 1981 joined together thirteen native organizations representing half the Amazonic native population. . . .

The worst hit victims of nuclear radiation in Sweden today are those who have never profited from its supposed benefits, who lived in and from nature. Dr Rosalie Bertell has been the voice of these and other victims of those who poison us for profit. She has exposed the corruption and the hidden value judgements of the nuclear establishment. Albert Schweitzer wrote in 1961: 'Only people who have never been present when a malformed child is born, people who have never heard its whimpering, never seen the sheer horror of the mother, only people who have no heart can support the madness of splitting the atom.' . . .

We have been requested by the Director of the Nobel Foundation not to refer to this Award as the Alternative Nobel Prize. The Right Livelihood Award was, of course, created and is presented at this time in this city because of a deeply felt and widely shared concern that the Nobel Prize had become a symbol of a system of values, and of a hierarchy of knowledge, which could no longer remain unchallenged if we are to save and heal this planet. We are of course happy to honour the request of the Nobel Foundation. Our Award for practical and exemplary work towards solving real problems facing humanity today speaks for itself. The letter from the Nobel Foundation concludes as follows: 'The Right Livelihood Awards have an important role to play in bringing attention to work which is important for the world and its future.'

1987

It was clear when these Awards were instituted that much work and knowledge vital to our well-being and survival was being ignored and rejected because it did not conform to the increasingly narrow vision of the world and its future which has come to dominate our lives, and which now threatens us all unless we can break its hold. The dangers we face are not just unfortunate side-effects but inherent in a vision of ever-increasing control over nature. In this world it is becoming increasingly difficult to do good: there are many more wrong than right livelihoods on offer. Even academics find it hard to ensure that they are not supporting the forces of destruction. Thus, Professor Charles Schwartz of the University of California recently saw no choice but to resign as teacher of advanced physics to avoid training scientists and engineers who would go on to produce sophisticated offensive weapons: 'social responsibility', he writes, 'means that a scientist to the best of his ability tries to foresee the consequences of his work for society.'

This Award was created to foster such a sense of social responsibility in us all. The support it has received has been inspiring and encouraging and has illustrated how great the need is. According to the United Nations Development Forum, this Award is today 'among the world's most prestigious'. It has become a beacon of light which reveals other paths, alternatives to the destructive and unwanted futures which are being predicted and prepared.

It would be naive to claim that we have reached a turning point, that we have yet achieved the critical mass to shift the loyalties of the few and stir the inertia of the many sufficiently to change the overall direction. But we have rekindled hope and shown what can be done, even with little or no outside support.

What are we facing? Earlier this year, the Brundtland Commission published its report entitled *Our Common Future*. The recommendations and conclusions it presents on the linked issues of environment and development are probably the most radical which can be hoped for from a body consisting of representatives of the global ruling elites. To quote one contributor to the report: 'We have seen the future and it does not work. We are headed for disaster.' Such warnings are of course not new. They echo Rachel Carson in 1962, U Thant in 1969, The Club of Rome in 1973, and President Carter's Commission in 1980, to name but a few. However, after examining the cures proposed in the Brundtland Report, one comes away with a feeling of Alice in Wonderland, for the underlying belief remains that the rich must get still richer in order to be able to help the poor and save our global environment! Basically, the failed and wasteful panacea of economic 'growth' is given another lease of life. The conclusion has to be that despite official hope expressed on all sides, no trends identifiable today, none of the programmes or policies deemed politically acceptable, offer any real hope of narrowing the gap between the rich and poor nations. . . .

We face an almost total failure of responsible leadership and nerve right through the political spectrum. While they may vent their pessimism in private, our 'opinion leaders' see it as their public duty to assure us that all is under control. . . . It was Sir Winston Churchill who said that politicians think of the next election while statesmen plan for the next generation. At this time, when the very existence of coming generations is in doubt, the need for statesmen and stateswomen is urgent if we are to avoid the breakdown to which present policies are inexorably leading. One of the first victims of such a breakdown will, of course, be democracy itself. For, in the ensuing chaos and calls for the 'strong man', the main choices on offer will be a new fascism proposing to sacrifice the poor majority in order to save the 'civilized' few, and a Stalinism of deprivation with strict rationing of resources.

This need not happen. In Gandhi's words, 'the world has enough for everyone's needs but not for everyone's greed.' The work of our Right Livelihood Award recipients in different areas confirms that these words still hold true—but for how much longer?

In this situation we are no longer prepared to accept as 'the best there is' societal structures and institutions which threaten the survival of our species, which reinforce and reward the worst instead of the best in human qualities, which make virtues out of almost every deadly sin, and elevate monetary wealth into an object of worship. . . .

Is it not high time that we took Gandhi's advice to cease to 'co-operate with our rulers when they displease us'? Is it not high time that we assert the realism of our values, and insist on policies that serve such values? It is inconceivable that we would poison our children and planet in order to provide a little extra material comfort for ourselves, that we would risk the future of mankind for the sake of political supremacy—for, to quote Frances Moore-Lappé, 'we have a strong sense of and need for fairness and co-operativeness that cannot be violated without great damage to the spirit.' Yet we have allowed our choices and responsibilities to become so abstracted and hidden behind a trust in the 'expertise' of others that we are prepared to allow those very policies to be carried out in our names! Why? . . .

There is a psychological resistance to recognize the immaterial limits of the scientific endeavour—the inability of the scientific worldview to provide us with the total picture. Science offers us no insights on normative values, final causes, existential meaning or intrinsic qualities. The problem is not so much that we (any longer) expect science to provide such answers or that science, despite all past promises, expects to be able to do so. The problem is that science claims to provide us with our sole reliable guide to 'objective' proven truth which, we are told, is the only 'real' truth and knowledge there is. The areas of knowledge science does not cover (including normative values, final causes, meaning and intrinsic quality) are thus neglected, delegitimized, doubted, relegated to the 'purely subjective'—the meaningless chance products of chemical reactions in our physical organs.

This, of course, is not really science but 'scientism', a new faith, both arrogant and intolerant. This faith raises us to masters of the universe while at the same time reducing us to automatons, programmed by chance, without will or freedom. Can there be any doubt about the effects of such a faith, indoctrinated daily throughout our media and educational systems by behaviourist psychologists and materialist philosophers and scientists, on our psyche and sense of self? Can anyone doubt that here lies a crucial reason for the feelings of impotence and disempowerment, for the 'unpreparedness to make connections' and face facts which seem to paralyse so many?

Scientism has made the 'enchanted garden' (Max Weber) of our world into a purposeless struggle of all against all. 'Those alive today are the descendants of winners' (Professor C. F. von Weizacker). Even altruism, we are told by sociobiology, is just genetic programming for species survival. But apparently, the programme does not include the information that we need an environment to survive, which is perhaps why Professor von Weizacker is now, at age eighty, having a nuclear shelter built under his house!

But this is no reason to let science get away any longer with the gigantic confidence-trick whereby it claims to be value-free and thus exempt from the ethics of the society in which it operates. Science demands to be allowed to research without such limitations and assures us in return that the results of this research will be for us all to judge according to our (personal and societal) values. But these same values have been delegitimized by scientism! As critics of nuclear power, genetic engineering, etc., know only too well, demands that the results of certain scientific research should not be used, will be dismissed by 'experts' as subjective emotionalism, irrationality, Luddism, etc.

What we need today is a science which respects and honours as legitimate and 'real' our deeply held values, our sense of meaning and our sense of the sacred. Otherwise, the anti-scientific backlash is bound to grow, coupled with a revival of intolerant fundamentalism—or with the nihilistic cynicism often found in the wake of Seveso and Chernobyl.

As Professor Willis Harman has written: 'The reality people encounter in the depth of their being is not the reality of Western science. It is not possible in the long run to build a well-working society on the basis of wrong assumptions about reality.' To those who reply that our intimations of reality are nothing but wishful thinking, that we need to base our sense of reality on what Bertrand Russell called the 'firm foundation of unyielding despair', the answer is threefold.

First, the scientific method of acquiring knowledge, the controlled experiment, can only find what is inferior to ourselves, for only that can be controlled. Thus, the truth will always be larger than what can be scientifically proved. Second, scientists like Francis Crick who want to 'explain all biology in terms of physics and chemistry' ignore the revolutionary developments in physics during this century. We know today that energy, not matter, is the primary level of reality. We know that laws of nature are not immutable but 'only' statistical probabilities. And we know that reductionism—explaining the superior by the inferior—is no longer a reliable investigative tool since the investigation itself interferes and changes the object investigated. The materialists in the 'soft' sciences have had the rug pulled from under their feet—but carry on most unscientifically as if nothing has happened! Third, vital parts of the modern worldview have—contrary to common belief—never been confirmed by science. The general theory of evolution, the belief that species evolve from each other by chance mutations, remains as unproven as when Darwin first proposed it. Scientific research in many areas over the past century has rendered it increasingly implausible. The innumerable 'missing links' required to fill the gaps remain missing, which to anyone not blinded by dogma suggests that they have never existed. Evolutionism is the great myth of our century. Myths are far from useless: they have great explanatory value. While in their creation myths our ancestors saw, and other people still see themselves as descended from gods, our scientific myth-makers have chosen to portray us as risen from the apes.

The purpose of these Awards is to help us overcome the self-imposed limitations of our shortsighted and narrow concept of reality. This Award highlights the work of those, whether scientists in the broadest sense or not, who have chosen to look further. We are not offering new dogmas but we are taking seriously the reality we encounter in the depth of our being: the knowledge that we are in some way part of a meaningful whole and that we are moral agents ultimately responsible for our actions as well as for our failures to act. In this spirit the Institute for Social Inventions in London has proposed the following Hippocratic Oath for scientists:

'I vow to practise my profession with conscience and dignity; I will strive to apply my skills only with the utmost respect for the well-being of humanity, the earth and all its species; I will not permit considerations of nationality, politics, prejudice or material advancement to interfere between my work and this duty to present and future generations; I make this oath solemnly, freely and upon my honour.'

We are here tonight to honour those whose lives exemplify the commitments of this oath.

Professor Johan Galtung's work is so prolific that I feel he must have secretly invented the seventy-two hour day! His contributions to peace research and peace education are without parallel. He is an original yet profoundly practical and precise visionary who has provided key ideas and theories in the fields of social science research, community action, ecology, and development—and in our search for values. He is a guiding light for social movements worldwide. He is a principled yet kind fighter and a true globalist. To quote one of his nominators: 'His critique is taken seriously even by those who always think they are right.'

Chipko is one of the most inspiring and broadly-based grassroots movements in the Third World. The activists (mostly women), poets and scientists of Chipko have taken great risks and fought many struggles to preserve their most precious heritage, the forests (which, in the words of Sunderlal Bahuguna, 'provide the poor with fruit, fuel, fodder, fertilizer and fibre'), from being cut down and turned into timber and paper for the affluent. At this time, when the tree-felling bans imposed by Mrs Indira Gandhi are again being threatened by greed and biased science, it is especially appropriate that we honour this movement whose slogan is 'ecology is permanent economy'.

Frances Moore-Lappé's pioneering work has revealed the true economic and political reasons for hunger in the world. She has also, with her co-workers, provided us with empowering knowledge about our own responsibilities and abilities to do something about this man-made tragedy. She has reminded us that this is a question of freedom, the freedom to construct a social order consonant with our most deeply held values. Her books range from bestsellers like *Diet for a Small Planet*, *Food First: Beyond the Myth of Scarcity* and *Aid as Obstacle* to the recent *What to Do if You Turn Off the TV*. Knowing that in the USA the average young person has seen 18,000 murders on TV by the time he or she reaches the age of eighteen, this book alone is worth an award!

Behind this Award lies the conviction that each one of us, whatever our situation or background, has an important contribution to make to the well-being of our planet. This Award is for those who, when the chance comes, put humanity first. There can hardly be a greater danger to us all and a greater obstacle to arms control than secret nuclear weapons programmes and further nuclear proliferation. It is curious that politicians who preach to us daily about the interdependence of our world, and about the common fate of humanity, still react with shock when someone acts accordingly. Today, there can be no treason except against humanity. How much longer will our legal systems ignore this simple truth?

When he discovered his country's nuclear weapons programme, conducted in collaboration with South Africa and France, Mordechai Vanunu put the interests of us all first, irrespective of the consequences for himself. He was kidnapped in breach of international law in Italy by Israeli agents. He is now imprisoned under inhuman conditions and facing the threat of a life sentence. His action is an inspiration and example to us all. This Award is also an expression of outrage at his treatment by the Israeli authorities. We call on the government of Israel to avoid further damage to its reputation and to drop all charges against Mordechai Vanunu. We call on all governments in the Middle East who have not yet signed the Nuclear Non-Proliferation Treaty to do so in order to facilitate a nuclear-free zone in the Middle East.

Professor Hans-Peter Dürr, physicist and interdisciplinarian, has investigated methods to make arms-control agreements safe against cheating. He has been involved in developing the concept of non-offensive defence, i.e. a military capacity structurally incapable of offensive actions. This theory has already entered the dialogue between the two German states and the two superpowers.

Professor Dürr has highlighted the responsibilities of scientists as teachers of the young. The Global Challenges Network, which he has created and towards which Award funds will go, is no ordinary military-to-civilian conversion project. Such projects usually start from the needs of the technologists rather than from the needs of the planet. Thus, instead of SDI we get 'civilian' space stations. The Global Challenges Network looks at the needs of the planet and its people and transforms these needs into concrete projects and challenges for the minds and energies of those now busy working on, to quote Professor Dürr, 'nonsense or worse'.

The Awards jury recognizes the essential contribution of the activists and thinkers of the peace movement in planting the seeds of sanity which have sprouted in some of our politicians this year. At the same time, we need to be clear that what has happened so far will be worthless unless followed by much more drastic action. If other weapons programmes continue, there will be yet more nuclear arms in the superpower arsenals in five years' time, despite the INF 'Double-Zero' agreement. We cannot allow ourselves to rest, as the peace movement did after the partial test-ban treaty in 1963....

Is there hope? When the Reverend Jesse Jackson was asked this question he replied: 'I get hope because in my lifetime I went to catch a bus with my mother and the sign above the driver's head said "Colored Seat From The Rear". That sign does not exist any more.' Even age-old prejudices can be changed within one generation.

The task facing us may be even more daunting, the time available even shorter. But just for that reason I believe we should take up this unprecedented challenge with joy and excitement, for to us has been entrusted nothing less than the saving of life on earth!

1988

The American writer William James once said that he was neither a pessimist nor an optimist but a 'possibilist'. I think that we in the Right Livelihood family—recipients, organizers, supporters—are possibilists, and tonight we are here to celebrate in this beautiful environment what is possible, what can be achieved by those who make serving the needs of the planet and its inhabitants their first priority.

The destruction of our natural environment, a crisis so serious that in the words of José Lutzenberger, 'Either we change our philosophy or we will really finish off life on the planet,' continues to get worse. To quote only a few examples: the most recent Environmental Action Programme of the EC Commission in Brussels concludes that the Community's environment 'continues to deteriorate'. After many years of highly publicized initiatives one would perhaps have expected that the situation, while still serious, was at least slowly improving. The fact that it is not even at a standstill but actually getting worse is, I feel, a declaration of political bankruptcy by those responsible.

In Holland the largest study ever of the environment, conducted by over fifty organizations under government auspices, was published earlier this week. Bearing in mind that this is a country with one of the strongest environmental lobbies in Europe, the conclusions are horrific, leaving, as one major newspaper put it, 'hardly a glimmer of hope for the future'. The groundwater is 'poisoned' and in a few years not a single healthy tree is expected to remain. The worst polluters are cars, intensive cattle ranching and the chemical industry.

A review last month in *The New York Times* of recent reports on the global environment and the Third World by the Worldwatch

Institute, the World Resources Institute and the International Institute for Environment and Development is headed 'The worst is yet to come'. *The International Herald Tribune* reported two days ago that 'The Mediterranean's pollution problems have only just begun.' A document handed to me a few days ago in Munich is headed: 'Climate predictions: increasingly catastrophic'. Another report published last week (*Die Zeit*, December 12, 1988) is headed 'Cars or life: the costs of automobile traffic can no longer be paid'. It comments, 'A technique which kills over a thousand children per annum would never stand a chance if one tried to introduce it tomorrow in full knowledge of the consequences.'

The Swedish environmental expert Björn Gillberg claims in his latest book that 60,000 Swedes die prematurely each year due to environmental pollution and that their life expectancy is shortened on average by fourteen years. In a similar study the respected US statistician Professor Jay Gould has concluded that there were 30,000 excess deaths in the USA in 1986 alone for which the Chernobyl radioactive cloud was the only plausible explanation. In some parts of Poland child mortality rates due to pollution are now far higher than in the Third World. As for animals and plants, it is expected that one fifth of all species may disappear over the next twenty years (Worldwatch Institute).

Enough! Indeed much more than enough. It is quite obvious that our system needs drastic reforms as urgently as does the Soviet Union, but this message seems not yet to have brought about any 'glasnost' (chemical companies are still allowed to withhold details of their waste releases as 'commercial secrets'), much less any 'perestroika'. It is still basically Business As Usual. The limits to growth are to be forcibly extended once more at great cost to our future. Western Europe is to become an 'internal market', where, to quote the British Ambassador to Bonn 'the economic reality will determine the political and social and not vice versa.' Already polluting industries in the EEC can save as much as sixty per cent of their production costs by moving to the member state with the most lax environmental regulations. Instead of technologies, products and trade appropriate to the needs of the planet, we are promised a market without a government where any attempts to replace the 'free' market economy by, for example, an ecological 'circular' economy based on re-use, recycling and reconditioning will no longer be possible, will be a breach of treaty.

Already the GATT free trade agreement is being used to further open up the Third World to the kind of trade which our previous Award-recipient Anwar Fazal has compared in its effects to an armed attack. Swedish authorities have decided that subsidizing urgently needed pollution control equipment for Poland is not permitted under GATT rules!

As for the Third World, the Vice-President of the World Bank, himself from a Third World country, declared in Berlin recently that the best way for 'the rich countries to help Africa was to stimulate their own economic growth by consuming more'! At the same meeting an International Monetary Fund economic spokesman declared that 'it is something of a puzzle [that] strong growth in the industrial countries has not affected the growth of underdeveloped countries.' The fact that last year the poor countries paid back more than $50 billion to the rich, more than the total they received in new grants and—mostly—loans was obviously not regarded as fitting into the puzzle.

Having been told by such authoritative sources that we should eat more in order to help the starving, no wonder many Westerners are confused. Living in an increasingly push-button society without time for roots, rest or relations, we register what would appear to be increasingly schizophrenic behaviour. Thus recent surveys show a large decline over the last few years in values involving altruistic activities and social concerns, and an increasing priority to making money—while other surveys find similarly large majorities in favour of greater equality and justice and less materialism. But I feel that this contradiction is only apparent. There is no willingness to make personal sacrifices seen, perhaps wrongly, as meaningless in the total context—but I believe there is a readiness to respond to a new paradigm, a new vision and plan which is on a par with the problems we face and seems to offer some real chance of turning off the road to disaster. But in a tragic miscalculation of the public mood, our present establishment does not appear to have the courage to propose what is needed, assuming that a public which objects to small changes cannot be persuaded to join in the implementation of a grand design. . . .

What I am talking about is a new perspective in which the whole is more than the sum of its parts. We need a new definition of wealth, of waste and of consumption. . . . What is required is an initiative of global solidarity and ecological responsibility, similar in scope and commitment to the initiatives which the international community took in 1945 to prevent another world war, to restore prosperity and prevent a repetition of the pre-war economic collapse. Such initiatives were the creation of the United Nations, the Bretton Woods agreement and the Marshall Plan. It is not my intention here to comment on the success and shortcomings of these initiatives, simply to illustrate what is needed: a tall order, perhaps, but by no means impossible.

Our recipients tonight represent corner-stones of this new world order. The work of the Rehabilitation and Research Centre for Torture Victims (RCT) aims to heal the survivors of torture, to educate about this crime against our humanity and dignity, to expose the perpetrators and their assistants so that they can be brought to justice. I feel it is high time that an international community worthy of its name institutes a total boycott of all regimes found guilty of practising torture. . . .

Wilderness is necessary for the growth of the spirit. Even those who live in the city feel the violence done to us by the theft and destruction of our forests, air, water and soil. How much more then must those suffer for whom the forest is their ancestral and spiritual home, as well as their livelihood? How much sharper must be the pain of witnessing the destruction of the forests for those who are still in tune with nature, whose identities cannot be separated from the trees? How must it feel to have not just your homes but your soul destroyed, cut down and sold for money?

That is what is happening in Sarawak. Harrison Ngau is in the forefront of the struggle against this sacrilege and crime. He should have been with us here tonight, but the authorities in Sarawak, by arresting his co-workers, made it impossible for him to come. We are shocked and I know you will all join me in expressing our protest against these actions, both to the Government of Malaysia and the

State Government of Sarawak. We urge that all those arrested be released immediately and unconditionally, and we urge that the wholesale destruction of the forests of Sarawak be halted now. We ask this not just for the sake of the peoples of the forest, not just for us all who are diminished by this destruction, but we ask it also in the interest of Malaysia, for no country can survive and prosper by destroying its heritage.

It is with great pleasure that I welcome here tonight Mr Mohamed Idris, Founder and President, and Mr Martin Khor, Vice-President of Sahabat Alam (Friends of the Earth) Malaysia.

The destruction of the forests to provide short-term enrichment for outsiders is not restricted to Sarawak. Much of what I have just said can also be applied to the Amazon where José Lutzenberger's work is based. He has been called the father of the environmental movement in Brazil. He is an indefatigable optimist who, since he left the chemical industry in protest against its policies, has taken up many causes and looked for the roots of the crises we face. He has achieved remarkable victories, not least in the massive reduction in the use in his home state of what he has termed agricultural poisons. He has written: 'We must analyse the technologies that we use, especially new technologies. We must ask ourselves who conceived them for what purpose, whom they are benefiting and whom do they harm and how will they integrate into the environment?' These are the crucial questions in the new paradigm we are creating.

John F. Charlewood Turner reminds us that housing and habitation are inseparable from community. He teaches that empowerment cannot be conferred but only found within ourselves, but that without access to resources and services, attempts to empower are a recipe for frustration. Doing more with less is the principle we all need to learn—which, of course, does not mean turning backwards but, on the contrary, using our resources with ever greater sophistication.

In this year when the refusal of the experts to be 'on tap' (available as and when needed) instead of 'on top' has brought abuse even on the head of Prince Charles, who was accused of being 'A Prince for the past' for daring to attack soulless modernism in architecture, it is very appropriate that we honour a man who, in many countries, has shown that architecture can serve as a catalyst to create a much higher quality of living and participation than the old paradigm admits to be possible....

1989

Welcome to our Tenth Award Presentation!

We began in a rented hall with an audience of thirty, so we have come a long way. To those who have joined me in the past decade and worked to make this Award what it is today I say a heartfelt 'Thank You!' Personally I can conceive of no other work which would have given greater pleasure than this: to be able to support and work with the most exciting individuals and most worthwhile initiatives in the world today. There is a project to describe the work of our recipients and nominees in a 'catalogue of hope'. Nothing could be more appropriate.

In 1942 H.G. Wells wrote:

> There will be no day of days then when a new world order comes into being. Step by step and here and there it will arrive, and, even as it comes into being it will develop fresh perspectives, discover unsuspected problems, and go on to new adventures.

Looking back over ten years of Right Livelihood Award recipients I think you will find that their steps lead in the same direction and their perspectives have much in common. These are not isolated initiatives but corner-stones of a sustainable global community. Next year we are going to bring many of our recipients together for a workshop with prominent politicians, international officials and media representatives in order to identify more closely the blocks which prevent us moving faster in the direction required. For, although this Award has received widespread recognition in many countries and given inspiration to millions, there can be no doubt that our world today is in a worse state than ten years ago. Even the pessimists failed to foresee the speed of destruction in many areas: 'Not even the scariest models predicted anything like the Antarctic (ozone) hole.' (Bill McKibben, *The End of Nature*)

The threat is global and immediate. 'The life-support systems of our planet are collapsing,' says not some alarmist but the director of UNEP, Dr Tolba. To give one example: Sierra Leone suffered a forty per cent reduction in food production because of increased cloud cover in the last two or three years. Meteorological experts working for the UN traced the increase of cloud cover, which was almost constant during the growing season, to Brazil. It was huge clouds of black smoke coming from the burning rainforests.

After yet another environmental conference, Robert Lamb, director of the UNEP-sponsored Television Trust for the Environment, wrote earlier this year:

> As someone involved full-time in the international environmental movement for more than a decade, I have attended dozens of meetings. This was the first time that I came away with the feeling that the problems are so overwhelming that rich and poor alike will have to awake and apply a common global strategy if a catastrophe is to be avoided. There is, however, just a ghost of a chance that public opinion, North and South, will reach a point where governments will have to tackle the root of environmental impoverishment. Something radical has to be done. . . .

Over the last year I have been told by men in positions of power that 'nobody told them' about the dangers of poisonous wastes, that building high chimneys was expected to solve the acid rain problem because 'nobody thought it would come down again'; that water pollution came as a surprise because 'water always used to clean itself' and so on. A theologian told me that the second Vatican Council said nothing on the environment because the problem was 'not known then'! . . .

But the establishment did not know because it did not want to hear such heresies against the prevailing dogma that science, technology and economic growth would automatically solve all problems, including those they create. . . .

Such shortsightedness is as disastrous as dogmas which are being swept aside in Eastern Europe . . .

Today the old ways no longer work, even by their own criteria. The ecological costs of economic growth are now estimated at twenty per cent in West Germany, but are increasing four times faster than the

GNP. The global climate is changing at ten to sixty times its natural rate due to human interference. This makes nonsense of much future planning and present activity. A group of scientists meeting in Austria in 1985 concluded:

> Many important economic and social decisions are being made today on major irrigation, hydro-power and other water projects, on agricultural land use . . . and coastal engineering projects; and on energy planning, all based on the assumptions about climate a number of decades into the future. Most such decisions assume that past climatical data, without modification, are a reliable guide to the future. This is no longer a good assumption.

However, there is hope! As our Award recipient Amory Lovins has pointed out, there are 'new technologies—albeit of a more mundane vernacular and "transparent" kind . . . such as insulation rather than fusion reactors and microelectronic motor control rather than solar power satellites [which] have proven far more powerful than anyone . . . thought possible. These relatively small, accessible and cheap technologies even seem powerful enough to solve the energy problem, the water problem, . . . and probably a good many other thorny problems to boot.' Lovins points out that it is we who are now the 'technological optimists' as opposed to conventional wisdom that clings desperately to the old gigantic failures.

I believe it is now time to go one step further and create an organizational structure which can end our war with nature, establish and implement everyone's legal right to a healthy environment, and support the transition to a sustainable world order. The environment as a global survival issue needs to be removed from the arena of day-to-day national politics. It cannot wait for the establishment of a world or European government—which are probably not desirable anyway.

Let us therefore take the issue, on which there is widespread agreement that national solutions are not possible, and create an institution with the political, financial and legal powers needed to restore and preserve our natural environment. Such an institution would not restrict 'freedom', but would provide the only possible framework within which human freedoms—including 'free markets' —can operate, without self-destructing.

Any nation or group of nations can initiate this Environment Authority by agreeing—by a referendum or parliamentary vote—to transfer the necessary powers. The authority's directorate would be elected by universal suffrage in all member countries. For practical reasons, each continent would have its own authority, which would delegate some of its members to the board of the World Environment Authority, which would have the final say in any dispute.

I do not believe this proposal is utopian. It is, like these Awards a decade ago, another idea whose time has come. It is time to 'go to scale' and implement on a larger level what we know works.

Our Honorary Award Recipient this year represents an inspiring step in that direction. Alternative economics, in order to have an impact, cannot just withdraw from our society but must aim to create a new mainstream. We honour the Seikatsu Club Consumers' Co-operative as a project which has successfully taken up this challenge, proving that it is possible to grow into an organization of over 500,000 members without losing the original vision—ecologically sustainable production, exchange and consumption, based on co-operation, self-management and abolition of waste.

It is no coincidence that three of the recipients tonight come from Ethiopia. We felt it was important to show how much good and successful work goes on in one African country alone. Our large institutions are happy to mouth platitudes about Third World self-help, but in practice Third World scientists often find themselves blocked unless they defer to the Western desire to control and define reality. Western science has produced miracles, but it has also 'grown fat, lazy and corrupt', to quote Professor James Lovelock, often regarded as Britain's foremost environmental scientist. It is time that Third World science is recognized and valued on its own terms.

Dr Melaku Worede has established the most successful programme of native seed conservation in Africa, and thus saved much of the continent's genetic seed resources from disappearance and oblivion. Together with Ethiopian farmers whose work and skills over millennia have created these resources, Dr Worede has established strategic seed reserves, in order to spread the knowledge of how to preserve plant genetic variety.

Dr Aklilu Lemma and Dr Legesse Wolde-Yohannes share our Award for the discovery of, and original and persistent research into, the parasite-killing properties of the endod plant, which can provide a cheap and natural way to eliminate bilharzia, one of the worst illnesses in the Third World, afflicting over 300 million people.

If you ask 'But then why haven't they received the Nobel Prize?', you are of course quite right. Drs Lemma and Wolde-Yohannes have had to work for twenty years to overcome the prejudices against Third World science in our elites, including governments, aid agencies, medical and international organizations. Resources have until now not been made available to spread the knowledge and use of the endod plant, thus forcing many millions to suffer and die from bilharzia needlessly. . . .

Survival International is a global campaigning and educating organization, which works with tribal peoples to secure their survival and land, their rights and self-determination. Survival International also works to educate us in the north on the importance of the tribal people of the world for our own survival. Largely due to their efforts, there is now a widespread awareness that tribal peoples are not to be preserved as museum relics of the past but as models of sustainable lifestyles in tune with nature.

Survival International is represented here tonight by its director, Stephen Corry, and by two representatives of one of their most urgent campaigns: Claudia Andujar and Davi Kopenawa Yanomami, who are working to save the Yanomami people from extinction and who were nominated for the Right Livelihood Award in their own right.

This is an Award for work in progress, a recognition of a unique and very effective organization, and at the same time an expression of our active support for the desperate struggle of a small people. . . .

Recently I came across these words by a political leader, now out of favour:

> If the present is obscure and the future opaque, if one does not know where one is, nor where one is going, then, naturally one asks the maximum for all and now.

These words summarize to me the spiritual malaise, the crisis of meaning and identity, which is so deep-seated in the industrialized world today. One aim of these Awards is to make the present less obscure, the future less opaque, so that one no longer needs to ask the maximum in order to feel secure. Ten years of Right Livelihood Awards have shown that there is another way and that it works.

I

Towards a Peaceful Planet

INTRODUCTION TO THE PROJECTS

Robert Jungk
1986
HONORARY AWARD
*'an indefatigable fighter for sane alternatives
and ecological awareness'*

ROBERT JUNGK was born in 1913 in Berlin, emigrated to Paris in 1933, where he made documentary films and studied at the Sorbonne, lived in Prague from 1936-38 where he published the anti-fascist paper, *Mondial Press*, and then fled to Switzerland when the Nazis entered Prague, staying there until 1945. Then as a freelance journalist he worked for several papers including *The Observer* in London, for which he covered the Nuremberg Trials.

During the 1950s he began to explore the themes which have dominated the rest of his life: the future, and peace and anti-nuclear activity. His first book was entitled *The Future has Already Begun*. In 1953 he founded the first Institute for Research into the Future, and in 1967 he co-founded with Johan Galtung the International Conference on Futurism, out of which emerged the World Federation for Future Research. He began to develop Future Workshops, in which people envision desirable futures and the means whereby they can achieve them, as a means of regaining power over their own lives and destinies. Then in 1987 he founded the International Futures Library in Salzburg, the first public library specializing in the collection, interdisciplinary networking and distribution of future-oriented information.

Robert Jungk
c/o International Futures Library
Imbergstrasse 2
A-5020 Salzburg, Austria

Mordechai Vanunu
1987
'for his courage and self-sacrifice in revealing the extent of Israel's nuclear weapons programme'

MORDECHAI VANUNU was born in Morocco and moved with his family to Israel in 1963. He did three years military service from 1971-74 when he was given an 'honourable discharge'. He became a technician at the Dimona nuclear plant in 1976. In October 1979 he began studies at Ben Gurion University, Be'er-Sheva, in philosophy and geography, graduating in 1984-85, when he became in succession a post-graduate student, assistant lecturer and external lecturer in philosophy.

At university Vanunu became increasingly politically active, calling for equal rights for Palestinians within the state and for the inclusion of Palestinians in negotiations for the establishment of an independent and separate state for Palestinians. He became increasingly disillusioned with Israel's military posture and opposed the 1982 Israeli invasion of Lebanon. In September 1986 Vanunu talked to *The Sunday Times* about the Dimona plant, revealing that Israel's nuclear capability was far greater than suspected; that Israel probably has a stockpile of 100-200 nuclear weapons; that it can make thermonuclear devices of greater power than atomic weapons; and that Israel also collaborated routinely with South Africa on nuclear matters. Soon after *The Sunday Times* article on October 5th, Vanunu went missing. In November it transpired that he was lured to Rome by Israel's Secret Service, then kidnapped and taken to Israel where he has been in gaol ever since. His trial for espionage and treason opened in Israel on August 30th 1987 under conditions of the most intense secrecy, and he was sentenced to eighteen years' imprisonment in March 1988.

Mordechai Vanunu
Ashkelon Prison
PO Box 17
Ashkelon, Israel

INTRODUCTION TO THE PROJECTS

Hans-Peter Dürr
1987
'for his profound critique of SDI *and work to convert high-technology to peaceful purposes'*

HANS-PETER DÜRR is fifty-nine years old, and the Director of the Heisenberg Institute of Physics at the Max Planck Institute of Physics and Astrophysics, and Professor of Physics at Ludwig Maximilian University, both in Munich, West Germany.

Dürr is a quintessential interdisciplinarian. He has been professionally active in the fields of energy policy (he has spoken and demonstrated against nuclear energy), science and responsibility and epistemology and philosophy, as well as in his specialties of elementary particle and nuclear physics. Recently he has also become very concerned about Third World economic and ecological matters, and since 1985 has been a member of the Board of Greenpeace Germany. But his main campaigning work in the 1980s has been on the theme of peace.

Hans-Peter Dürr
Max-Planck-Institut für Physik und Astrophysik
Fohringer Ring 6
D-8000 Munchen 40
West Germany

Johan Galtung
1987
HONORARY AWARD
*'for his systematic and multidisciplinary study
of the conditions which can lead to peace'*

JOHAN GALTUNG was born to Norwegian parents in 1930. He has had an international academic career spanning thirty years, five continents, a dozen major positions and over thirty Visiting Professorships, fifty books and more than 1,000 published monographs.

Turning to the social sciences after initial research as a mathematician, Galtung published his influential *Theory and Methods of Social Research* (Allen & Unwin) in 1967. In 1959 he set up the International Peace Research Institute in Oslo, the first institute of its kind to make a mark in the academic world, and was its Director for ten years. In Oslo too he founded the *Journal of Peace Research* in 1964 and edited it until 1974. He was Professor of Conflict and Peace Research at the University of Oslo from 1966-77, during which period he also helped to found the Inter-University Centre in Dubrovnik, Yugoslavia, as a meeting place for East and West, and was for four years its First Director-General. High ranking university positions followed in succeeding years, interspersed with consultancies to the whole range of UN agencies: UNESCO, UNCTAD, WHO, ILO, FAO, UNU, UNEP, UNIDO, UNDP and UNITAR. Some of the subjects in which he held a Visiting Professorship in 1986 were international economics at Sichuan University, China; world politics of peace and war at Princeton University, USA; international studies at Duke University, USA; and Peace Studies at Chuo University, Japan. He is currently Professor of Peace Studies at the University of Hawaii, USA.

Johan Galtung
Institute of Peace
University of Hawaii
Honolulu, Hawaii 96822
USA

Making the Future
ROBERT JUNGK

December 1986

IN THE TALK, DEAR JAKOB VON UEXKULL, which you gave during the 1984 ceremonies, you spoke of the longing for a very different future from the one we are now creating. It is precisely for such a different future that the previous recipients have been working. But who are 'we'? Can we also be equated with the planners and builders of the technocracy? When I reflected I realized that this statement is unfortunately true: through our work, our tax money, our consumption we have become captive supporters of the system, which pretends to create a wonderful future but in reality destroys all hope of one. We have become wheels in an all pervasive machine of destruction, which works against us, our children and their children.

However, there is a worldwide movement of resistance, which tries to liberate us from this new tyranny. It is growing. Not fast enough, but at a steady pace. It is strong, but by no means strong enough. Maybe something is lacking. To say NO is essential, but not sufficient, because negation is always dependent on that which it negates. A good doctor will not only fight a disease. He will try to stimulate the healing forces in the body.

Therefore, if the new movements for real change are going to succeed they will have to turn the 'longing for a very different future' into a loud and powerful 'Yes'. We need radically new concepts, concrete examples of more human relationships between people and just as importantly between people and the earth. These glimpses of another 'tomorrow' should be made visible even today.

On one of the walls in my workroom hangs a woodcut. It shows a half-naked man who tries to gather together the glowing debris of an exploded sun to form a new heavenly light. I feel very close to this figure, because for years I have been collecting bits of bright news: information about hopeful beginnings, encouraging activities, reports about imaginative individuals and groups, who try out new ways and do not give up.

The public does not hear enough of these seeds of a more sensible and hopeful future. Maybe that has to do with the way that the media look at reality. For them 'good news is no news'. Sensational events such as crises and catastrophes seem more interesting to them than the attempts to live more simply, to work more meaningfully, to help one another and to create spiritual wealth rather than accumulating material goods.

Future Research—at least in its first phase—did not listen to these faint signals of a new 'Zeitgeist'. What was heard was the bigger voice of the technocrats: more, faster, stronger, higher, bigger. The roots of this violent futurism can be found in the 'think tanks' of the American military and industrial establishments. Even before the Second World War ended the RAND Corporation was founded by the US Air Force in order to study and prepare future armed conflicts. Shortly thereafter 'Stanford Research Institute' came into being and in the area known today as Silicon Valley. Here plans and products for control of the future world market were developed.

Only now do we begin to understand the importance of these first attempts to invent and manipulate the future rather than letting it happen. Continuing in the tradition of Los Alamos and other weapons laboratories they devised blueprints for complex and extremely powerful systems. These differed significantly from the earlier inventions of the industrial age: their impact must be measured in decades, centuries or even thousands of years. Even more horrifying is the fact that the damage they may cause can be irreversible.

So today we have to live with the possibility of a man-made apocalypse. The sudden and final end of our species—alas—has become a realistic vision. The senseless extinction of numerous plants and animals portends what may be our own fate. For short term economic or political gains we risk damages which may last a hundred times longer or perhaps forever.

In the late sixties a new kind of futurist emerged. Their guiding star was PEACE not war. They spoke of 'old fashioned' ideas such as human scale organization, humility, beauty and the inescapable fact that man and woman belong to nature. It was no coincidence that it was the peace researcher Johan Galtung and the Quaker James Wellesley-Wesley, who organized the first worldwide conference on future research in 1967. I am proud of the fact that I was able to work with them at the very beginnings of this new movement.

We want to overcome the deadly crisis of our civilization by developing strategies for survival. Not competition, but co-operation, not exploitation of people and nature, but the care and protection of the creation and its creatures was to be our way into the unknown world of tomorrow. Such a vision lacks the grandiose and macabre fascination of technocratic plans. They don't have the sensational touch that gets the attention of the media and their consumers, who are out for thrills. And we should not underestimate the seductive power of that risky adventure.

Despite all that, the alternative culture, with its attempts to create a pluralistic decentralised future by choosing a soft path instead of the aggressive technology highway, gains momentum. Year by year the number of projects, whose participants think differently, work and live differently, increases. Recently I attended a conference in Hamburg, where more than 2,000 individuals from over seventy self-help organizations met to discuss their efforts towards a more liveable future.

Among the activities they described were: efforts to regenerate urban slums, the revival of neighbourhood democracy, new ways to heal industrial illnesses, the creation of meaningful and useful work, the development of alternative energy by citizen groups, the promotion of female culture, establishing spaces for children's activities, inspiring cultural creation at all levels and many other positive endeavours.

Such alternative networks, seedbeds of a new culture, have been growing in all industrial nations during the last few decades. Their members are not waiting for the 'big day' of sudden change. They are starting here and now to build convincing models of peaceful existence. The effects of their activities can be likened to acupuncture for the social body. Already now they are becoming like antibodies within the afflicted system. If it were not for them some paralysis would grip many people of good will, who are close to despair.

These self-help groups not only help themselves, but they give hope to many others. They do not give in to resignation, realizing that this would open the road to the destroyer of mankind.

A doubt persists: are these pioneers of a more human and ecological future not too weak? Can they really overcome the enormous power of the entrenched establishment? I would contest that. In times of instability even small inputs of new quality can dramatically alter quantitative superiority. In an epoch of mounting crises people, who can offer possible solutions, have greatly increased chances of influencing the course of events.

Such problem solving takes place in the 'future workshops' which have sprung up in a number of countries, especially in Germany and Denmark. In these new grassroots groups concerned citizens get together in order to develop their own visions of the future. They want to enter the political process early enough in its initial stages while new ideas are being generated. This way they can become subjects rather than objects, planners rather than victims taking part in the shaping of their destiny.

The people in power are much weaker than they appear and the citizen movements much stronger than they themselves realize. As defenders of the unborn, as protectors of the earth, as pioneers of peaceful relationships they create new beginnings in the middle of the old. We should give them more than our sympathy. They urgently need our active and lasting support. The alternative future is already here. The prize of honour you have given me is an encouragement for us all, who know: OTHER WAYS ARE POSSIBLE. I thank you, we thank you with all our hearts.

Conscientious Objection and Nuclear Secrecy

MORDECHAI VANUNU

December 1987

THE PASSIVE ACCEPTANCE AND COMPLACENCY with regard to the existence of nuclear weapons anywhere on earth is the disease of society today. This is the disease that we are going to pass on to future generations, together with the nuclear weapons themselves and the nuclear installations. However, more and more people are becoming aware of the danger which the very existence of such weapons entails, a realization which marks the beginning of efforts to put an end to all nuclear arms, not to mention preventing their further distribution to other states.

Nuclear arms are human artefacts—men made these destructive means for suicide. But man can also change his mind, can make another choice, can stop this production. Not everything which can be manufactured must be manufactured. If governments made the mistake and chose to develop these deadly weapons over the past forty years, at least today many of their citizens are consciously opposed to the production and possession of such weapons. This opposition is growing in many countries in both East and West, and now also in the Third World. This movement is gathering momentum and has actually influenced the governments of the United States and the Soviet Union to try and eliminate nuclear missiles in Europe. This shows us where the governments went wrong in the first place: they took it upon themselves to produce vast quantities of nuclear arms and now the people are seeking to turn back to an era before the constant threat of a world-wide nuclear holocaust.

I am speaking today not in my own name only, but in the name of many ordinary people who have been taking part in this struggle against nuclear arms. Some of these people have also been imprisoned, in Britain, Germany and the United States and other countries. This struggle is not only a legitimate one—it is a moral, inescapable struggle.

What we who are opposed to nuclear arms are saying is this: 'It is not we who broke the law, who violated human and civil rights, but the governments which chose to create the greatest threat to human life that ever existed.' Never in human history was there such a threat to the very existence of mankind, and to all forms of life on earth. In fact, up to this moment no one has been able to predict all the effects of nuclear war, of that global disaster. It is known that not only will vast numbers of civilians be killed and maimed, but that there will be all kinds of ecological effects afterwards. Carl Sagan has spoken of a 'climatic catastrophe' which will follow upon nuclear war. Scientists all over the world are prophesying disaster if nuclear war ever breaks out.

Man cannot continue to ignore this danger, and no government, not even the most democratic, can force us to live under this threat. No state in the world can offer any kind of security against this menace of a nuclear holocaust, or guarantee to prevent it. The only thing they can do is to keep producing more of these weapons and place them in the hands of human beings. But this is not the kind of weapon which is controllable by human reason. Human beings are apt to make mistakes, human judgement can fail. Just as there is no perfect man, so there is no perfect leader. Yet in handling nuclear weapons there must be no mistake, no failure. We have all seen the outcome of a 'minor' failure in the nuclear plant at Chernobyl—not even a bomb, only a technical failure, but the disaster was huge and spread throughout Europe. Neither borders nor governments could stop it from spreading. So this threat does not hang over a particular country, but over all people, everywhere.

We are united by the awareness of this threat to our lives, to the life of children and old people and to nature itself. There will be no future after a nuclear disaster, and it will be impossible to correct it and turn back. Correction and turning back can only be achieved now, today, by destroying all nuclear arms everywhere in the world, by returning to the pre-nuclear era. This must be done, despite the policies of the world powers and the various governments. We are the sane civilians, we want to live—and we say to the governments, 'No nuclear arms!' We must make it plain to them that we shall never accept the existence of nuclear weapons.

The first step in the struggle against nuclear arms is to know, to understand, to recognize the issue. This is why I am in prison, this is why many people are in prisons all over the world. We must arouse people and warn them. Ours is the role of the angry prophet. In the first stage we must make all citizens everywhere aware of what their governments are doing or hiding with regard to nuclear weapons. Even if we cannot stop it, at least we must know about it, warn people about it, give them information about it, and in this way increase the number of people who resist. This is the way to mental health. A person who refuses to accept the existence of this suicidal type of weapon is a mentally healthy person, who does not want to be expelled from this earth as Adam was expelled from the Garden of Eden.

There is an especially urgent danger that more and more countries, notably in the Third World, are now developing nuclear arms of their own. Some of them are poor underdeveloped countries which are supported by Western governments. The potential for the development of such weapons is becoming more feasible all the time. The possibility that every state on earth will produce nuclear arms is terrifying. Already now there are enough nuclear missiles to destroy the world many times over. What will happen when many more countries will have them? The danger to us all will be increased, the risk of a nuclear war breaking out will be multiplied. Who knows what the leadership of every one of these countries is capable of, or what the political conditions in each of them can lead to? There are many local and regional conflicts in the world, and who can guarantee that they will not result in the deployment of nuclear weapons?

This proliferation of nuclear weapons may lead the world powers to approach the problem in a sane way. As more and more countries aspire to possess them, the world powers are losing control of the situation. This should worry them sufficiently and lead them—the governments of the United States and the USSR—to begin to think like human beings. Only direct supervision and control, only a broad and determined action, will prevent the circle of nuclear arms from widening further, and eventually lead to the elimination of these weapons everywhere.

To start with, they will have to create nuclear-free zones, based on agreements between the local states and on international supervision of all nuclear installations. The Middle East could be such a nuclear-free zone, because it has known many wars and conflicts and because the world powers are also involved in the region, thus increasing the danger of a nuclear outbreak. The world powers must intervene to prevent proliferation and to impose international supervision, otherwise the momentum of nuclearization will increase throughout the region.

There is no such thing as a nuclear deterrence—certainly not in the Middle East. Israel and the Arab states must be made to conclude a treaty to eliminate nuclear weapons in their region. Perhaps such a treaty will also lead the citizens of these countries to get together and promote a peace process and eventually lead to good neighbourly relations between them. It is an illusion to believe that nuclear weapons can be defensive. Nuclear weapons are a means of extermination, and cannot protect any country, including Israel.

A state which lives in fear of destruction must not threaten the whole world with annihilation. The people who experienced a holocaust must not bring a holocaust upon another people. Moreover, a nuclear war is worse than the holocaust of World War II. A single bomb can eliminate a whole city, and everything that lives in it, and even the very infra-structure of the city. The effects could last for hundreds of years—and yet it is so easy to do. That is the real holocaust. That is the real enemy, and not as Israelis imagine.

Unfortunately, there is as yet no general awareness of the dangers of nuclear weapons in the Middle East. Many still think of them as a lifeline, whereas in fact they are a hanging rope, and not just for us but for all mankind. The nuclear threat resembles radioactive radiation, in that it is invisible yet deadly. Will people only realize this when it is too late?

Every nuclear installation creates an immediate problem, namely, radioactive waste. This is a major ecological danger. No country, not even the United States, has found a solution to the problem of radioactive waste. These substances will remain dangerous for thousands of years, and we are leaving this deadly inheritance to the generations which will follow us. By what right are we leaving them these wastes, with all the dangers that they entail and the problems and costs of dealing with them?

I am now in prison and cannot speak freely, as I would like. I can only express myself indirectly. But everyone knows what I did and what was my purpose and intention. It was a positive act, and it succeeded in part. I contributed my share to the struggle against the nuclear danger. Above all, I succeeded in raising the issue and all its implications even in this region, which is in a state of conflict. I stood up against the proliferation of nuclear weapons in the Middle East.

The movement for nuclear disarmament must spread everywhere, because nuclear arms are spreading everywhere. Organizations must be established in every country on earth, to awaken people's consciousness to the danger. The organizations which have been active in Britain, in Germany, in New Zealand, have done a great deal, not only for their own countries, but for the whole world. The steps taken by New Zealand with regard to nuclear arms—going so far as to ban the entry of ships bearing nuclear missiles into its territorial waters—should be adopted by other countries. Only thus can we safeguard the future of the world. I wish to continue to act against the proliferation of nuclear arms, because anyone who knows what is happening and is aware of the issue cannot evade it and must not remain silent. This is especially true for me now, because of what I have suffered in prison and will continue to suffer. Despite the suffering I am sure of myself and confident that I have acted properly in the face of the terrible danger of nuclear war which hangs over this region. The goal is worth the self-sacrifice, and I know that many other people in many countries are doing the same thing. There will certainly be more and more people who will resist their governments' policies in relation to the issue of nuclear weapons.

Any country which manufactures and stocks nuclear weapons is first of all endangering its own citizens. This is why the citizens must confront their government and warn it that it has no right to expose them to this danger. Because, in effect, the citizens are being held hostage by their own government, just as if they have been hijacked and deprived of their freedom and threatened. Therefore, when a man is held hostage by his own government the least he can do is resist it and its nuclear policy. He must do this in self-defence, which is a basic civil right. Indeed, when governments develop nuclear weapons without the consent of their citizens—and this is true in most cases—they are violating the basic rights of their citizens, the basic right not to live under constant threat of annihilation.

This issue raises many questions, among them the primary question: Is any government qualified and authorized to produce such weapons? The United States developed its atom bomb during World War II, without giving much thought to the long-term effects and consequences—and the rest of the world has followed it to this day.

But is it to be taken for granted that governments have the right to do so? I believe that it was a mistake to make nuclear weapons in the first place, and a mistake to go on producing them. Now at last many citizens are beginning to question their governments' right to manufacture these weapons. Is this what governments were elected to do? If, at the outset, governments themselves were not fully aware of the implications, neither were their citizens. Now there is more information, more people know what the dangers are, and resistance is becoming more widespread. The more information becomes available on the subject, the greater will be the resistance. It is essential, therefore, to spread information as much as possible, to uncover more and more so-called nuclear secrets which governments wish to keep from the people. These are not secrets of security or defence, because nuclear weapons are not defensive weapons, and they endanger the citizen rather than protect him. Nuclear secrecy in itself is the enemy, and the citizens must know all about it.

Thus the only guarantee of survival is peace; the only guarantee for the future development of human civilization is peace, true peace—not only between states but between peoples, religions, and races. Human fraternity is the basis for this, and it begins with the elimination of barriers and frontiers. Nothing is so safe as peace, peace in which every state respects the right of the other, in which all co-operate to develop and advance the region they inhabit. That is what is needed in the Middle East today—a true peace between all the states in the area. Only peace will secure the future of the state of Israel. Just as Israel made peace with Egypt, so it must continue. Without peace there will be no security. Without peace even nuclear bombs will be of no use to any state. So if there is one thing that must be done, must be continuously worked for, it is the political effort to achieve peace, by means of negotiations, by means of direct contact and mutual recognition, a recognition of the other's right to exist, without the threat of war.

Israel has attained the summit of its military might [*Here some words were crossed out by the Israeli authorities*] but all this military might has not produced a condition of peace and security. There is still a state of war and men, women and children are being killed. So we have no choice but to proceed to invest more in the effort to achieve peace. War only leads to war and then eventually they may lead to a nuclear disaster. But perhaps the danger of nuclear disaster will drive the people in the region to take a chance on peace. Forty years of war in our region are more than enough. There is a limit to what people can stand. Now they are weary of war and want to live in peace, to feel that there is no more danger, that their children are not destined to be soldiers, that children are brought into this world to live in it, not to fight.

It is not true that war is an inevitable state of affairs—on the contrary, what is essential is peace. Eventually the leaders will also reach this conclusion. Therefore what I have done is not only in connection with the nuclear issue, but with the problem of the Middle East conflict as a whole. It is meant to point out the danger, to call out a warning against the fearful consequences of the existing situation to the entire world.

Few people have an opportunity to confront this issue as I have done. It is a question of listening to your conscience, to the voice of Jesus telling you, 'Yes, this is what you must do—you must sacrifice yourself, your personal freedom.' In the nuclear era, and with the nuclear threat hanging over us, the answer to this challenge is obvious. One may not evade such a responsibility, one must accept the mission in order to warn against the danger [*Here some words were crossed out by the censors*].

I acted on behalf of my countrymen and for the sake of all mankind. What happened in Hiroshima was not just another page in history, but a major landmark, pointing out the crossroads for mankind—this may be the fate of our planet if we do not resist it. In our region I took the first step to show that the danger is real and immediate. Now the rest must be done also, to lead back from the nuclear menace to a treaty of nuclear disarmament. We must not keep silent, we must not avoid confronting reality. Others must continue what I started. Let us act together for the sake of this region, and love one another. If we fail to do so, death will destroy us all. It is up to us to act. We must not leave the nuclear issue to the politicians. The citizens themselves must put their governments on notice and tell them, 'No nuclear weapons! No nuclear holocaust!' We are living in an age of uncertainty, living with our future in doubt. Every year that passes like this may be a year taken from our future. If we comprehend the danger and fight against it, we may yet save ourselves and all mankind.

The Global Challenge of Disarmament
HANS-PETER DÜRR
December 1987

As a scientist I am here to argue against the terrible misuse of science. Not only huge sums of money but also an increasing part of our intellectual resources are wasted today in designing more sophisticated, more insidious, more powerful weapons and counterweapons. And not only that. While trying to curb or halt this fatal military build-up, even more of our best scientists and our highly motivated people are being drawn into the military machinery although their original motivation was just the opposite, namely to keep out of it. They often find themselves in the awkward position of being forced to participate in a game they do not want to play at all.

Many of us, for example, were involved and are still involved in the debate on the Strategic Defence Initiative (SDI), in particular concerning its technical feasibility, its stability, and its political implications. Our efforts, as far as we can see, unfortunately have done nothing to stop the project. On the contrary, sometimes it appears that they have even stimulated its advocates and their well-paid experts to suggest new and improved proposals. Or sometimes one gets the impression that such phantoms are merely created to keep us busy and to divert our attention and the attention of the public from even nastier developments. For my part, I hate to be misused in this way.

Surely we all have more important and more reasonable things to do than to argue against these crazy and dangerous projects. We certainly all would prefer to use our precious time and energy on something constructive, something which will contribute to fostering co-operation, compassion and friendliness among people, and to preserving the life-sustaining capacity and the beauty of the earth.

Of course, I realize that we cannot abandon military questions altogether since the most acute threat to peace at the moment would seem to be the mad and accelerating arms race and the resultant destabilization of the world military situation. The most likely cause of a nuclear war, it appears, is not the failure of mutual deterrence—what sane person who desires his own survival and that of his species would willingly start a nuclear war? The most likely cause of a nuclear war will be the inability of military-technical structures to handle a political crisis. The weapon arsenals of East and West are like closely linked parts of a single huge nuclear reactor. Contrary, however, to our so-called 'safe' civilian reactors, this weapon reactor is constructed in such a way that in case of failure it does not switch off but will escalate to full destruction.

Therefore, unfortunately, we must continue to pay careful attention to military-technical developments. We have to make great efforts to break the dynamics of the arms race and to find ways to improve crisis stability generally. We have to realize that this cannot be achieved by a straightforward approach. The arms race is not simply caused by some vicious people—although I know, of course, that there are such people and even very powerful ones—but is predominantly the result of an eigendynamics which strongly constrains the system of armaments to its present pernicious course.

Our inability to stop the arms race is partly connected to our lack of understanding of the dynamics of the process. We are used to looking at phenomena as static and do not give enough attention to their development in time as it results from their integration into a more general causal structure. In particular strong feed-back mechanisms cause the system to get out of control and to proceed on its own destructive path—much as a microphone-loudspeaker system starts shrieking if the amplifier is turned up too much. Static considerations dominate our thinking and are in many instances the basis of our decisions. And this has its reasons. In the case of closed systems or approximately decoupled systems, static thinking allows us—by using our past experience—to anticipate and to predict the future behaviour of the system to a certain extent, and in this way also offers the means to control and regulate the system. For open systems or systems strongly coupled to their surroundings, this is no longer possible. These systems will not respond easily to exterior controls but are predominantly regulated by inner constraints.

Static thinking, for example, may prompt us, as in the case of SDI, to construct a huge defence system in outer space, a few hundred kilometres away from the earth's surface, to destroy all attacking missiles or their warheads on their trajectories (just as we would try to avert the strokes of an opponent's sword by means of a sturdy shield). Dynamic thinking will teach us, however, that the construction of such a defence system in space will necessarily generate countermeasures from the opponent which may completely defeat its original purpose. In judging the ultimate effectiveness of a defence system as envisaged by SDI, it does not suffice to consider the technical feasibility of any particular lay-out of such a system but to take fully into account all possible reactions to it. As a technical undertaking, therefore, SDI cannot be compared, for example, with the Apollo project, the landing of man on the moon, because the moon did not ward off the human effort, it did not shoot back. If we consider the extreme vulnerability of space-based sensors as required by a missile defence system and, on the other hand, the relative robustness of missiles and their nuclear warheads, we easily understand the disadvantage of the defender. Each technical breakthrough in constructing a tighter umbrella against nuclear missiles will at the same time provide even better means to punch holes in it. This only indicates the uselessness of SDI for the purpose for which it is advertised for the public, namely for providing protection against nuclear missiles. If this were all we could simply ignore it as a threat. The real danger of SDI, however, lies in the fact that it will bring weapons into space and therefore carry the arms race into a new dimension, jeopardizing among other things the stabilizing function of satellites. In addition, the huge SDI research and development programme will in its wake generate other terrible weapons.

Similarly, a nuclear test stop appears to change very little from a static point of view because it will not reduce the nuclear arsenals nor prevent their further increase. A comprehensive nuclear test ban would, however, constitute a clear signal that both sides do not intend to base their security forever on deterrence and the principle of mutual assured destruction. The test ban, therefore, would start a process in which all parties have to think about new and better ways of stabilizing the military situation.

The idea of military equilibrium in the sense of exact weapon arsenals is another example of the inadequacy of static thinking. Because of the basic asymmetries in the opponent's situations, the continuous development of weapon technology, deficient perception of the opponent's strength and intentions and the ambivalence of weapons regarding their defensive and offensive use, each side will only feel safe if it actually is stronger than the other side, a situation which will necessarily lead to an arms race. One way to break up this vicious circle would be to restructure one's defence forces in such a way as to make them incapable of attack without diminishing their effectiveness in defence. Sufficient defence combined with structural inability for attack, a non-offensive defence posture, appears to be a salient stability principle on the level of conventional arms. The military on both sides are doubtful about the possibility of a reliable non-offensive defence because in many instances attack is still assumed to be the best defence. There are some indications, however, that these doubts are being slowly eroded, as reflected in official statements in East and West during the last year. Many people quite naturally do not like the idea of restructuring military forces. They clearly prefer disarmament. So do I. The question is how to achieve disarmament, and whether it is possible to go directly from the arms race dynamics to a disarmament dynamics without certain transition stages, as for example, transarmament. In considering non-offensive defence as a transition posture one should not underrate the psychological factor. The general atmosphere in which negotiations proceed is dramatically improved if the potential opponents do not limit the discussion merely to establishing some parity in their over-kill arsenals, but begin considering ways and means to reduce their mutually perceived threat.

All this work on military-political and military-technical problems, I wish to stress again, is, I believe, extremely important and will, unfortunately, also require the active participation of scientists and technicians to some extent, but it is also evident that peace in its real sense can never be achieved by military measures or technical fixes. Military-technical measures at best will only lengthen the fuse. They may stretch the time for solving the underlying basic problems or, better, they may provide the time necessary for a learning process which directs our attention towards these problems. But lengthening the fuse only helps if the time gained is used to disarm the charge. It is, indeed, high time that we focus our attention on the real problems which are threatening all of us, in fact life itself, on this planet.

What are the real global challenges we are facing? We may give them different priorities, but we all recognize the following urgent problems:

▷ How can we harmonize a rapidly growing industrialization with our vulnerable environment on which we vitally depend?

▷ How can we secure in the long run our energy needs without threatening present and future generations with deadly risks?

▷ How can we change the state of affairs where, despite increasing productivity, a growing part of a rapidly growing human population is harassed by poverty, hunger and disease?

▷ How can we achieve more justice on earth? How can we reorganize the world economy so that the ample fruits of the earth and the abundant goods of human production do not accumulate where there is affluence but go where they are really needed?

▷ How can we prevent the suppression of individual freedom and development by dictatorial power and narrow-minded bureaucracies? What must we do to ensure that human rights, in the wake of poverty, hunger and unemployment, do not in fact degenerate to the privilege of only a few?

All these problems can develop into worldwide catastrophes soon: if we do not address them resolutely they all will jeopardize our security. They all will lead to unrest, uprisings and wars. And war in our time can mean omnicide. It does not suffice to prevent wars but, to prepare for a lasting peace, we have to do much more: we have to approach and tackle all the urgent problems and this without delay. Why don't we all—East and West, North and South—join hands to meet the great challenges confronting humankind? Because these problems concern us all and equally. Why should it not be possible to make these really urgent global problems of our time for once the main focus of a comprehensive research and development programme, the object of a global challenges initiative? And this directly and explicitly instead of hoping that their solution will more or less accidentally occur as a by-product or spin-off of a military-technical super-project like SDI.

Of course, I realize that such an undertaking would be colossal and extremely complicated, its goal is utopian. But should we therefore discard it as a goal? Is the goal envisaged by SDI not totally utopian, too, and aren't there some people—people, in fact, who pride themselves on being particularly realistic and pragmatic—who seriously intend to realize it? And this just with SDI, which obviously only constitutes another step up in the arms race. Of course, I realize that SDI was proposed by the President of the USA and, most important, quite a lot of money was announced. It may also be that the task of an initiative which tries to solve the urgent global problems is even more complicated than SDI because these global problems are not only of a scientific-technical nature. But such an initiative would be so much more worthwhile, so much more reasonable and so much more suitable for general consensus. The brightest and most enlightened people in all fields, on all levels of society, and from all geographical areas should be persuaded to join this effort—and I believe they also could be persuaded and motivated if they are approached in an appropriate way. Many are waiting to be asked.

How do we want to implement this crazy plan? How can we ever hope to transform this vision into solid reality?

We find ourselves faced with the common dilemma:

▷ On the one hand, the large number of closely interrelated problems requires us to step back and consider the full picture and to understand their structure and their dynamic behaviour.
▷ On the other hand, our limited ability to devise solutions and to effect change by concrete actions forces us to narrow our attention and resources to a few problems at a time.

We can, however, escape this basic dilemma by proceeding on two different levels at once: a general conceptual level and a concrete practical level.

On the general conceptual level, we should identify the basic global problems and illuminate their structure, their causes, their interdependencies and their possible developments. Study projects like Global 2000, for example, may give us a good start. Such a general survey will certainly leave us with a great many questions, but with very few answers.

Looking for solutions to these problems, we must consider some basic questions like:

▷ How do we want to live?
▷ What are our priorities?
▷ What should be our means of resolving conflicts?

These are very difficult questions which do not allow definite answers. In considering these questions we should not hesitate to formulate and project utopian goals to serve as general orientations for practical action. Because of the basic difference of the present historical situation, past experiences will only have limited value. We should therefore be prepared to enter and explore completely new territory and engage in a new way of thinking.

Because of the varying cultural and intellectual backgrounds of the world's population, these questions will obviously not receive unanimous answers. But they should be asked anyway. And this should be done not to throw up new walls between people but, on the contrary, to find common ground. Solutions, in general, do not require unanimity, but only compatibility. Despite the great variety in people's opinions and lifestyles, we should not jump to the conclusion that there is no common ground between them. In fact:

▷ We all have to proceed on the basis of a limited earth with limited resources in its accessible crust;
▷ We all are convinced that every man, woman and child on this earth should have the opportunity to lead a decent life;
▷ We all are vitally interested in the survival of humankind and the biosphere.

These goals should actually suffice to establish a solid basis for common actions.

The main question, of course, will be how to translate our distant and utopian goals into manageable first-step actions. On the basis of our general conceptual considerations a multitude of very specific and concrete problems will emerge that could serve as starting points for practical solutions to the larger problems. These first-step problems constitute the concrete practical level. They deserve our full attention. Because the concrete problems are highly specific and vary greatly in quality, their investigation and solution require the active participation of very many people on various individual and institutional levels. Where do we find these people and groups?

I am convinced that there are many people who would be willing to participate in this effort if we only can offer them a chance. There is already a multitude of organizations and groups involved in problems connected with peace, ecology, energy and Third World issues. Some are frustrated by their long and seemingly unsuccessful struggle because there is barely any official support for them, to say the least, and the tasks exceed their abilities and strength. These tasks are really too difficult for all of us. It is, therefore, important that we not only indicate utopian goals for orientation, but that we intensively think about possible ways which may lead to them and point out the very first tiny steps which have to be taken. It will also be important to provide room for creative action for everybody who wants to get involved and wants to participate.

But how do we go about implementing these ideas?

As a first step towards this general goal, some of us, beginning this year, have started a network which we called the Global Challenges Network (GCN) for the purpose of developing an international net of projects and groups who will co-operate in a differentiated way in tackling the urgent global problems, the global challenges. Such an idea, of course, is not new. It has been tried in the past in various ways without ever really succeeding. It is a crazy idea, a pure utopia. In our case, at least for the moment, it shall simply serve the purpose of a general framework in which a study group will be formed, an International Science and Technology Study Group—ISAT-SG. The study group shall consist of competent and knowledgeable men and women from various countries and professions with theoretical and practical experience, i.e. people with so-called T-intelligence.

T-intelligence here shall mean an intelligence which can best be symbolized by the capital letter 'T' combining a vertical bar, indicating depth and detailed expertise, and a horizontal bar, indicating a global, holistic view and broad experience in which special knowledge is harmoniously embedded. The study group shall in particular have the tasks of:

▷ pointing out the most urgent global problems;
▷ structuring these problems according to
  – topics and problems areas
  – the methods of approach for implementing and solving these problems
  – the availability of material and intellectual resources for their realization;
▷ breaking them down into smaller, simpler and more accessible subproblems and projects;
▷ defining selection criteria for assigning priorities to the subproblems regarding, for example,
  – their general relevance
  – their urgency
  – the feasibility of possible solutions
  – the possibility for global co-operation
  – the transferability to different geographical regions
  – the number of people involved
  – the size of the necessary financial investments
  – the time period for realization
  – their symbolic power and their novelty value;
▷ figuring out and suggesting practical entries to possible solution;
▷ identifying worldwide scientific, technical, cultural, etc. expert-type and grassroot-type capacities on a global and local level, competent for working on detailed solutions;
▷ finding sources of political and financial support.

To prepare the International Science and Technology Study Group about forty people from various international organizations experienced in the relevant problem areas met in Feldafing near Munich in July 1987. The task we set ourselves was to make a first review of the whole problem areas, to advance suggestions for candidates for the first round of the study group and to develop ideas about a data-based computer network which can facilitate the co-operation of the study group members and, later on, can also provide important information on projects, methods of approach, and people and groups involved.

Some people have made the criticism that the study group has been given the name of a Science and Technology Study Group, which seems to imply that we assume that all the global problems are only scientific and technical problems and therefore should be solved by scientists and technicians. This point, indeed, requires some clarification. Obviously only some parts of the problems which threaten humankind are of a technical nature or are such that we can hope to tackle them by scientific methods. In fact, our whole method of approach in seeking solutions to problems, namely by cutting them up into a large number of smaller sub-problems, may prove quite inappropriate. There, indeed, will be a great many problems which we just cannot get hold of with analytical methods.

Clearly a successful solution to all the difficult problems will require ultimately a change in our perception, a change in the ways we communicate and deal with each other and with our environment. It will require a new way of thinking, as it is often emphasized today, which after all may not be really a new way of thinking but in a sense resembles more our old traditional thinking which had a more modest image of man concerning his abilities and his role in the world. It will require a whole cultural process in which science and technology will only play a minor part and which, therefore, has to be promoted by all members of our culture and not by scientists alone. Still, to rectify consequences of the scientific-technical revolution, to tame the unleashed scientific-technical monster, the active participation of scientists appears mandatory. Because of their closeness and active involvement in this sweeping scientific-technical process with its deadly threats to humankind, many of the scientists today are deeply concerned about the situation and therefore feel called upon to offer their assistance. Their excellent international contacts and their extensive experience in constructive international collaborations—which even survived the darkest periods of the Cold

War—in a way predestines them as pioneers for the global process of co-operation. This process, therefore, has to be initiated by scientists but it will only succeed if people from all areas of our cultural life eventually get involved and carry it on. Scientists can serve as catalysts in this difficult process.

What does this imply?

Among the huge number of unresolved problems which emerge from our general considerations there will certainly be a large number which could be successfully approached by scientific and technical methods. These problems, I assume, will be the simpler ones and easiest to be approached and solved in the present world political situation, and therefore they offer themselves as starting projects.

There is another consideration which favours this approach. At present, in many countries proposals are being advanced for non-military initiatives to stimulate national economies, as for example the civilian European technology programme EUREKA and the European space research programme. There is also a widespread readiness to embark on East-West scientific and technological ventures despite the present hostile political climate. Such joint programmes should be strongly encouraged. Their general orientation, however, should be changed, because most of these proposals focus on particular new technologies rather than actual needs. In view of the rising costs of research and development in new fields (e.g. space research), it seems to me very important to reverse the customary approach. Instead of starting with a specific technology—which in many cases is initiated and inspired by military deliberations—and only later considering its civilian application, we should decide which problems we want to solve, define priorities according to their urgency and social importance, devise possible solutions, and then push appropriate research and development programmes accordingly.

The driving force behind many of the vast technological developments and innovations at present is not predominantly generated by a desire to offer man more extended chances in life or a better quality of life, but rather, I am afraid, by the zeal to increase the profits of an economic élite and to strengthen the power of a few over the many. Our daily experience seems to indicate that basic necessities of our life get more and more subjugated to techniques and material conditions rather than conversely, that technique and material conditions are used and developed for the solution of the difficult and really challenging problems of our time. It is interesting to observe that some of this élite—or even many of them—have already recognized the absurdity of our present situation and their catastrophe-bound course but because of external constraints, in particular to remain competitive on the national and international market, they see no way to get out of this madhouse without jeopardizing their existence. Mere insight into the various phenomena, therefore, does not suffice. We have to know a lot more about the inner dynamics of the system. Using this knowledge we have to find ways to break up the eigendynamics of the process and try to recover its flexibility, and with it our ability to steer and to control this process in the small domain where we live. Because our quality of life, as we perceive it, depends decisively on the size of space around us into which our creative powers can expand.

In this context we should realize that the evolution of the universe and the evolution of life on earth points in the direction of generating structures of higher and higher order, in the sense that these structures get more and more diverse and differentiated. This concept of order is somewhat in contrast to our everyday conception of order which considers, for example, the arrangement of atoms in a salt crystal to be more orderly than the atoms, let us say, in the DNA-double-helix, or, as another example, the carbon atoms in a single graphite crystal more orderly than the ones spread out as printing ink of various letters in a book of poetry. These examples show that only the one who reads will be able to perceive this lofty stage of order, this tremendous bulk of information hidden in the seemingly chaotic and arbitrary arrangement of a DNA-supermolecule or a book. Due to our limited understanding of nature, and the very limited ways of non-destructively manipulating her as compared to her inherent potentialities, technical developments, in their effort to maximize certain preselected properties valuable to us in a certain context, automatically will sacrifice other potentialities and the flexibility in other ranges. Nature does not try to maximize certain features irrespective of others, but tries to optimize their systems in a high-dimensional space of a huge number of options. The extension to higher dimensions opens up many evolutionary paths and increases stability and robustness. A DNA-supermolecule with its atoms monotonously and orderly rearranged along a single line may be more easily memorized and synthetically reproduced, but such a molecule would have completely lost its property as a code for steering and controlling the growth of a particular living creature. Decentralization and differentiation of structures will avoid the dullness and inherent destructiveness of monocultures, and provide the ground for the conservation and the generation of higher organizational forms.

But will the economic forces which apparently favour centralization, standardization and power accumulation not prevent any change of the present course? I do not think that this has to be so. An undertaking which tries to solve urgent global problems and to meet fundamental human needs rather than to chase continuously after new expensive techniques should actually prove more successful in a competitive situation, at least in the long run. Because all these problems will not simply die away but even get worse in time, some of them or all of them one day will stand full-size at our doorsteps and the people being prepared will have great advantages.

Perhaps the solution of the urgent global problems may not require quite such extreme technologies and high-tech as is necessary for tackling space research or SDI. Therefore, some people may be afraid that these problems will not be intellectually ambitious enough to catch the imagination and the enthusiasm of our scientists and technicians, and that they are not glamorous enough to nourish their vanity. Some scientists will certainly prefer to see their name attached to a star rather than to a waste utilization plant. We should recognize, however, that in view of the increasing threats and dangers to man, and in view of the speed with which we are racing to catastrophe, many people—and in particular young people—during the last decades have become strongly motivated to devote their work and their intellectual and moral energies to real human needs.

The International Science and Technology Study Group will, so I hope, point out a large number of different ways to approach global problems. It may also name and motivate competent people and groups of people to actually tackle them. All these different project groups which, depending on their particular tasks, will be in close or loose contact with each other and appropriately will join their efforts, would become the nodes in an international network, the Global Challenges Network or whatever we may eventually call it.

The conditions for building such an international network have improved dramatically during the last year. In the Soviet Union an International Foundation for Survival and Humanity was formed at the occasion of the International Peace Forum in Moscow in February 1987 by an international initiative group. Preparations to establish similar foundations in the USA, Sweden and other countries are on the way. There are excellent prospects at the moment that all these initiatives will eventually fuse into one global network in the very near future.

The various projects of Global Challenges Network will offer an excellent chance for close co-operation between East and West, in particular if we first concentrate on the problems of ecology and energy, where both sides are 'in the same boat'. Actually, a common interest exists for all our global problems, in particular also the military security problem. But the present political polarization, unfortunately, prevents many from realizing this. Therefore, attempts to fully resolve East-West problems will probably have to wait for a more auspicious political climate. This climate, however, can be greatly improved by joint ventures. We should start where we face the fewest obstacles. Co-operation in projects of common interest is the best way to build up mutual trust and confidence.

Peace Studies: Inspiration, Objectives, Achievement
JOHAN GALTUNG
December 1987

THE HISTORY OF PEACE STUDIES is, of course, a collective story, and it did not start in the 1950s—it is as old as humankind, as is the history of war studies, the study of war with warlike means. I guess the same applies to what today is called security studies, or the study of peace with warlike means, balance of power, balance of terror, deterrence through credible retaliation, and so on. Needless to say, that is not what peace studies are about. Peace studies are about peace with peaceful means, and I think it is generally accepted today that this has to be done in a holistic manner and with a global perspective, or, in more narrow terminology, through interdisciplinary and international perspectives.

I am often asked to say something about what brought me into this field. And it is quickly done: I owe it to three factors that loomed large when I was eighteen years old:

1. The Norwegian military and the institution of general conscription—to that little letter calling on me to serve in the armed forces of my country, at that time a NATO country, tied to one of the world's superpowers.
2. A very gentle, very sweet and very cultured father who had spoilt his only son for many years by always permitting me, even encouraging me, to ask that essential question, 'Why?', patiently trying to answer increasingly inquisitive enquiries.
3. The German occupation of Norway 1940-45, when we experienced soldiers with 'Gott mit uns' on their buckles (that same god they praise in Norway), leaving all Norwegians with two basic questions: what do we do to avoid anything like this

in the future, and, if we should be occupied again, is there a nonviolent, peaceful alternative to armed resistance? These two questions had added urgency for me because my father, one night in February 1944, was hauled off to a concentration camp from which he returned when the war was over, but marked by the experience.

So, nothing mysterious; simple sources of inspiration. When after three years of reflection on this I finally applied for status as a conscientious objector, I added as an afterthought that I would like to devote my life to peace studies, a term with hardly any meaning, including to me. However, by that time I had read quite a lot of literature that was not included in the mathematics curriculum I was supposed to study, including work by Gandhi, Freud, Einstein, etc. Out of sheer tenacity I nevertheless finally got my Ph.D. in mathematics. In the process I had become a very poor mathematician indeed, working on some mathematics that has never been heard about since, but a reasonably good peace researcher.

So much for inspiration. What about objectives? They are two, neither of them modest.

The first is the goal and it is exactly the same as for the peace movement: the abolition of war as a social institution. It sounds naïve, doesn't it? Well, so did the talk of those people who wanted to abolish slavery as a social institution—they were all told they were working against something very dark and mysterious called human nature. And yet they somehow succeeded; there are still elements of slavery and colonialism around, but not as legitimate social institutions, and that is a very basic point. For this to happen alternatives have to be found; and this is a very difficult point. In one sense plantation colonialism was the alternative to slavery, and neo-colonialism of the Third World the alternative to colonialism, both highly unacceptable. A peace researcher should be mindful of this point lest we end up with an equally unacceptable alternative to war, something that could qualify for the fourth and missing category above: War with peaceful means. A world government referring to its belligerence as police action, for instance?

The second objective is more like a means: to make peace studies academically acceptable. Here great progress has been made from the first beginnings in the late 1950s. Peace studies are taught at more than one hundred universities and colleges in the US, for instance. More importantly: we have something to teach, a great pool of knowledge on which the two important qualifiers, holistic and global, may increasingly be applied—not to the single piece of research, but to the totality.

In short, we have something to teach. Hence, it is very obvious what the next step is—degrees in peace studies, from as many universities in the world as possible. We need thousands of them, holding an MPS, (Master of Peace Studies), serving international organizations, governmental and nongovernmental, transnational corporations with a conscience—and they exist, governments with a conscience—and they also exist, voluntary organizations such as religious bodies and trade unions and, above all, that new major peace actor in the world scene: the municipality. And, if this list is still not convincing: think of the many positions that have to be opened to teach peace studies at the level of elementary schools and high schools!

The time has passed when the only job opportunities for peace studies graduates was more peace studies—but there is still much work to do. The School of Peace Studies at Bradford University has blazed an important trail in the same direction; so does the Section for Peace Research at Uppsala University, and the Peace and Conflict Studies programme at the University of California, Berkeley, just to mention some. But much, much more is needed and on a much larger scale; and this work forms a major task for other institutions, like the University of Hawaii for the Pacific Hemisphere, or the Austrian Peace Research and Education Institute for the European region. For this can only be done at places with an imaginative, creative leadership.

Let me try at this point to give a very brief summary of where we stand in peace studies. If peace is the reduction of violence, like the abolition of war and related phenomena, then we have to start with a better conceptualization of violence than the word 'war' alone. I have found it useful to distinguish between three types of violence:

▷ direct violence, often expressed as military power, usually killing quickly, and intended to do so;
▷ structural violence, often expressed as economic power, usually unintended, killing slowly;
▷ cultural violence, often expressed as cultural power, legitimizing the other two types of power, telling those who wield power that they have a right to do so, even a duty—for instance because the victims of direct and/or structural power are pagans, savages, atheists, kulaks, communists, or whatever.

I have come to see the first two as relatively simple. There are ways of reducing large-scale abuses of military and economic power, such as deterrence through non-provocative, defensive defence as an alternative to the long range weapons systems underlying deterrence by retaliation (which to the other side looks suspiciously like an offensive capability). And there are ways of reducing large-scale abuses of economic power through economic self-reliance, at the regional (Third World), national and local levels, using indigenous production factors rather than becoming dependent on somebody else ('internalizing externalities' and 'sharing externalities equally' would be the more technical formulations). Rethinking military and economic science!

But cultural violence, in the form of religions and ideologies that announce themselves as the only valid faiths, for the whole world and in addition with a Chosen People appointed to spread that faith to others, not only as a right, but as a duty, is more difficult to handle. Here we are touching a cornerstone of many people's identity; a lie, for sure, but as Ibsen said, take that lie away from the average person and you remove his happiness at the same time. Would that not itself be violence? A key problem for peace studies, certainly not solved by rejecting it.

I let that do as examples, hopefully appetizing. Peace appeals to the heart; studies to the brain. Both are needed, indeed indispensable. But equally indispensable is a valid link between brain and heart. And that, in a nutshell, is what peace studies and peace practice are all about.

2

The Rights of the Planet:

Good Ecology and Sustainable Development

INTRODUCTION TO THE PROJECTS

Ladakh Ecological Development Group (LEDEG)
1986

'for preserving the traditional culture and values of Ladakh against the onslaught of tourism and development'

Ladakh is a high-altitude desert on the western edge of the Tibetan plateau, part of the Indian state of Jammu and Kashmir, with a population of 120,000 in an area of 40,000 acres. Until 1975 it was almost completely isolated from Western trends and influence and had developed a rich, harmonious, stable and sustainable culture. Then tourism began to develop, bringing with it many of the ills of Western maldevelopment.

The Ladakh Project was founded in 1978 by Helena Norberg-Hodge to affirm the beauty and value of many aspects of Ladakh's ancient culture and to enable Ladakhis to improve their standard of living without destroying it; to make information available about the costs of the Western development model and to stress the similarities between many traditional practices in Ladakh and powerful new trends emerging in industrial countries, trends for example towards holistic health, natural food, organic agriculture and environmental concern.

Inspired by the Ladakh Project, and with its ongoing involvement and participation, LEDEG was founded by Ladakhis in 1983 with the same aims, encouraging Ladakhi people to create for themselves a future which fully meets the particular needs of their unique environment and culture.

The Secretary
Ladakh Ecological Development Group
Leh, 194101 Ladakh,
Jammu & Kashmir, India

Ladakh Project
21 Victoria Square, Clifton
Bristol BS8 4ES UK

The Chipko Movement
1987
'for its dedication to the conservation, restoration and ecologically-sound use of India's natural resources'

The forests of India are a critical resource for the subsistence of rural peoples throughout the country. As these forests have been increasingly felled for commerce and industry, Indian villagers have sought to protect their livelihoods through the Gandhian method of satyagraha non-violent resistance. In the 1970s and 1980s this resistance to the destruction of forests spread throughout India and became organized and known as the Chipko Movement.

Chipko Information Centre
PO Silyara via Ghansali
Tehri-Garhwal
UP, India 249155

Frances Moore-Lappé
Institute for Food and Development Policy
1987
'for revealing the political and economic causes of world hunger and how citizens can help to remedy them'

FRANCES MOORE-LAPPÉ was born in 1944 and came to public notice in 1971 with the publication of her *Diet for a Small Planet*, which has sold three million copies and been translated into French, German, Swedish, Japanese and Spanish.

In 1975 Moore-Lappé and Joseph Collins founded the Institute for Food and Development Policy (IFDP), which has since become internationally recognized for addressing the political and economic roots of world hunger and demonstrating how ordinary citizens can effectively help to end hunger.

Frances Moore-Lappé
Institute for Food and Development Policy
145 Ninth Street
San Francisco, CA 94103 USA

INTRODUCTION TO THE PROJECTS

José Lutzenberger
1988
'for his protection and conservation of the natural environment in Brazil and abroad'

JOSÉ LUTZENBERGER is sixty-two years old and is a Brazilian agronomist and engineer who worked for fifteen years for the chemical company BASF, but left in 1972 to start a vigorous and successful campaign against the over-use of agrochemicals. In the ten years after 1978 the use of such chemicals in his home state of Rio Grande del Sul has fallen by more than seventy per cent, which largely reflects his tireless work with local farmers and their association on 'regenerative agriculture'—the process of increasing the fertility of the soil through food production rather than the reverse. His work in this field has made him an acknowledged expert in soil science and organic fertilizers.

> José Lutzenberger
> Jacintho Gomes 39
> Porto Allegre
> RS9000 Brazil

Dr Melaku Worede
Plant Genetic Resources Centre
1989
'for his leadership towards a sustainable end to hunger by building one of the finest seed conservation centres in the world'

MELAKU WOREDE was born in Ethiopia in 1936 and has spent his life as an agronomist. After obtaining a Ph.D. in Agronomy (Genetics and Breeding) from the University of Nebraska in the US, he returned to Ethiopia and became involved in the planning of the Plant Genetic Resources Centre in Addis Ababa, of which he became Director in 1979, a post he still holds.

Ethiopia is one of the world's eight 'Vavilov Centres' noted for their great genetic diversity. This diversity is now under great threat from drought and modern farming methods. It is this biodiversity that Worede has sought to preserve and, going beyond this, to establish 'Strategic Seed Reserves' of traditional seed varieties that can be released to farmers for planting in times of drought when no other seeds are likely to thrive. In only a few years Worede and his staff have collected and safely stored a large amount of the genetic diversity of Ethiopia, so crucial to the agriculture of both Ethiopia and elsewhere. In the process, Worede has established not just the finest African facility, but one of the world's premier genetic conservation systems, where collection, conservation, documentation and use are integrated and co-ordinated under one roof.

Worede has built this institution exclusively with Ethiopians, training a whole new generation of plant breeders and geneticists in his home country. Moreover, he regards his local farmers as crucial partners in his conservation programme. In 1988 he launched the world's first systematic effort to support farmer-conserver-breeders with scientific information. With his colleagues at the gene bank and NGO support, Worede is restoring the local seeds to farmers and helping them to improve their evaluation of each season's harvest. Together they expect to increase annual yields by as much as five per cent each year.

Dr Melaku Worede, Director
Plant Genetic Resources Centre
PO Box 30726
Addis Ababa, Ethiopia
Tel: 251 1 180381

Appropriate Technology and Co-operative Culture in Ladakh

HELENA NORBERG-HODGE

December 1986

LADAKH IS A REGION in the northernmost part of India, lying north of the Himalayas, in the ranges bordering the Tibetan plateau. Until a few years ago, Ladakh was one of the very few places that had not been affected by the Western monoculture that had spread across the entire world; in fact, when I arrived there in 1975, life in the villages was as it had been for eight hundred years.

Despite its size (about forty thousand square miles), the population of Ladakh is only about one hundred and twenty thousand. Situated in the rainshadow of the Great Himalaya, it is a desert. Coupled with the high altitude, the result is an extremely harsh environment. From the beginning, therefore, the people have had to recognize that Nature has limits. In order to survive, they have developed traditions which enable them to live in harmony and balance with their natural environment—traditions which have maintained a stable population and prevented division of the land. The villages are self-sufficient—in many ways, models of Gandhian village democracies. In the alluvial fans of glacial melt-water streams, fields have been carved out of the desert, and irrigated. Staple crops are barley and wheat, and there is some grazing in high summer pastures for sheep, cows, yaks and 'dzo' (a hybrid of yak and cow). Ladakh has survived only by using its natural resources carefully, and never abusing them. There is absolutely no waste. The scarce trees—apricot, willow and poplar—are not used for fuel, despite the forty degrees of frost experienced in the Ladakhi winters. Rather, they are carefully tended, and their wood used only for construction or for musical instruments and tools. Dried animal dung is used for fuel, and human nightsoil as fertilizer: every house has a composting lavatory, and all 'waste' is recycled.

STANDARD OF LIVING

Through this careful use of resources, and by living closely in tune with the earth, the Ladakhis have managed to create a society with a relatively high standard of living. When one considers the constraints—the extreme scarcity of resources, the harsh and difficult environment and the fact that these people have used only 'Stone Age' technologies—their success is quite remarkable. Many people in the West would think it impossible even to survive in these circumstances. Yet the Ladakhis have more than survived; they have actually prospered. Virtually everyone is well-fed and healthy, and people even have produced enough excess with which to buy luxuries—fine brocade, jewellery and precious metals. What is more, this is all done in a relatively short working season. During the four months of summer, they meet all their basic needs of food, clothing and shelter. Even most of the weaving and spinning is done during this period. For eight months, there is a lot of leisure, with weddings which last two weeks, monastery festivals, storytelling, and music-making. Water has to be fetched and all animals fed, but this is just about all the work that one has during winter. Even in the peak of the working season, during the harvest, when people work for eighteen hours in a day, play is mixed with work. The whole family and friends are together in the fields; everyone—from great-grandparents to great grandchildren—helps and sings together. And there is always time to stop to chat and drink the local barley beer.

In terms of quality of life, the Ladakhis do rather better than one would expect from a pre-industrial society. There is a high level of co-operation between all members of the society, with little distinction between rich and poor, male and female, old and young. Roles are very flexible. Women do some jobs more than men and vice versa; but rigid distinctions are rare. There is very little specialization, and as a result, work is rarely monotonous and boring. Everyone knows how to plant, how to build houses, how to make music, how to spin. Crime of any sort is so uncommon as to be virtually nonexistent. You can walk alone at night, without the slightest worry. Even after getting drunk at parties people do not get aggressive.

Co-operation, rather than competition, lies at the foundation of Ladakhi society. It can be seen in all spheres of life; from the sharing of household tasks and rotational shepherding to the interaction between children. One interesting observation that I have made in this regard is that in Ladakh, children are never segregated into age groups. Instead, they spend their entire lives constantly surrounded by and interacting with people of all ages. The implications of this are enormous. Ladakhi children benefit from the help and support that older companions can give. Imagine a room filled with thirty one-year olds. None of them can walk properly; they are all struggling to gain their balance. How can one possibly help the other? Now imagine another room filled with people aged one to thirty. Imagine the difference. Imagine how different two individuals would become if their lives were spent in such different contexts and societies. This is but one of many reasons why I have had to dramatically alter my beliefs about human nature. In seeing the extreme differences between this completely pre-industrial and traditional society and industrial ones, I have had to conclude that one's social environment does affect and mould one to a tremendous extent: especially when it comes to such important characteristics as co-operation and aggression.

Perhaps the most important characteristic of the Ladakhis that forced me to rethink my beliefs about human nature is the remarkable *joie de vivre* of the people. At first I thought that the Ladakhis smiled a lot and appeared very happy, but surely underneath they were just like all human beings—with their problems and feelings of jealousy, anger and depression. But after some years of living with them, I started realizing that all that laughter was connected to a deep sense of peace and contentedness. Even more dramatically, as Ladakh started changing because of outside influences and modernization, it became very clear that the people in the 'modern' sector were beginning to develop the same signs of depression, restlessness, anger and aggression that I was familiar with from the West. Observing the individual Ladakhis change, as the technologies, economic pressures and education—in other words, their society—changed, was the most convincing evidence that human beings are very dramatically affected by social pressures. And traditional Ladakh has proved to me that it is possible to have a society which encourages co-operation and happiness rather than the opposite.

Ladakh is a predominantly Buddhist area, with a religious tradition going back almost two thousand years. The signs of Buddhism are everywhere. Chortens, symbolic of the basic teaching, dot the landscape. Every village has its monastery, every path and pass its cairn of prayer stones. But interestingly enough, there are also substantial numbers of Muslims, and in the capital, Leh—small pockets of Hindus, Christians and Sikhs; and remarkably all these religious groups co-exist peacefully.

It seems to me that all the great religions of the world have as their central core the unity—the interdependence—of all life. Buddhism, in its very philosophical way, places particular emphasis on the limited nature of the world of the senses (in direct antithesis to science which acts as if only that which can be perceived, isolated and objectively measured has any value). While things appear to have their own absolute, separate existence, and they actually do on the level of the senses, the most important underlying, transcendent truth is the interdependence of all things and the unity of the individual and the universe. If you meditate, if you reflect, this 'oneness' becomes apparent, and the illusion of separateness disappears.

This world view based on interdependence goes hand-in-hand with a way of life where everything seems to be connected, where people are aware of their dependence both on other people and on the earth, and where these connections are clearly visible and harmonious.

The contrast with our own society is striking. In the West, the rise of science has been paralleled by the decline of beliefs (and everywhere in the world, one can see that the modernization process has the same effect), and our world view, which insists that everything is separate from everything else, has created a world where people are separated from one another and from the earth.

I may have given you the impression that Ladakh is heaven on earth, that everything is perfect. This is far from true; in fact, nothing is perfect. There would always be ways of 'improving' things, whether in health, agriculture or architecture. In particular, the physical discomfort—mainly due to the extreme cold—is a real problem. However, it is important not to lose sight of the totality. Overall, there is perhaps not a single place in the world today that can compare with Ladakh. It is a peaceful, settled and sustainable society.

TRAGIC CHANGE

Since Ladakh is in many ways a model society, it is tragic to see young people starting to reject their culture as primitive and filthy. As I said earlier, Ladakh had been almost completely unaffected by the West when I arrived in 1975. (This was a rare situation: the Indians of the Americas, the Aborigines and many other indigenous peoples were dramatically affected by our European culture many centuries ago.) In 1960, a road between Kashmir and Ladakh had been completed, but the area was still largely unchanged, and—for strategic reasons—out-of-bounds to foreigners. But suddenly, in the autumn of 1974, it was thrown open to tourism.

When I arrived, as the first Westerner that many people had ever seen, the young and old alike were proud of their society, proud of what they had, and they considered themselves wealthy. I remember very well going to the village of Hemis Shukpachan, as the guest of a friend, Norboo, and because his village was particularly beautiful, with large palatial houses, I asked him, out of curiosity, to show me the poorest house in the village. He thought for a bit: 'We don't have any poor houses,' he said. That was nine years ago. Last year I overheard Norboo saying to a tourist, 'Oh, if you could only help us Ladakhis, we're so poor.'

This is the heartbreaking change that I am now seeing: the people's perception of themselves is changing dramatically. Because of the very distorted picture of the outside world that they are gaining through contact with tourists, young people are beginning to think of themselves as poor and deprived. Since Ladakh first opened up, there has been an annual invasion of wealthy Westerners (as many as fifteen thousand a year). The Westerners are rich, and can travel thousands of miles for pleasure. They come for a few days, and spend perhaps £100 a day. In a subsistence economy, where basic needs are met without money, this is as if Martians would come to Bristol and spend £50,000 a day. £100 is what a family in Ladakh might spend in a year (only using money for luxuries, as they do).

The impact on the young is disastrous; they suddenly feel that their parents and grandparents must be stupid to be working and getting dirty, when everyone else is having such a good time—spending vast quantities of money travelling and not working. For Ladakhis, work means physical work; the notion of stress resulting from mental work is unknown. So they get the impression that if you are modern, you simply don't work; the machines do it for you. Understandably, the effect is that they try to prove that they are not part of this primitive bunch of farmers, but part of the new elegant modern world, with jeans, sunglasses, a radio, a motorbike. It is not that the blue-jeans (often uncomfortable) are intrinsically of interest; they are symbols of the modern world. Similarly, cinema films give the impression that racing around in sports cars shooting people—that violence—is modern and admirable.

Suddenly, being a Ladakhi is just not good enough; everything is done in order to seem modern. Tragically, this means leaving the village for the capital, where the money can be earned that will buy all the trappings of the modern world. It follows that basic needs can no longer be met locally. 'Modern' clothing must be imported; in modern concrete houses, one must eat modern food; one must eat imported white rice and flour. These modern concrete houses are the product of the changes in attitude that I have just described, coupled with conventional development practices. The traditional architecture is satisfying to the eye; it reflects the environment around it—it uses local materials: adobe, stone and wood. The flat-roofed houses reflect the climate in a rainless land. The new buildings in Leh are a complete antithesis—the same concrete boxes, symbols of the global monoculture which one finds in Florence, Peking and Los Angeles. They are ugly and alien and not adapted to any environment. We are told that they are more economical, yet the materials have to be bought, and in the case of Ladakh, dragged across three high Himalayan passes, two days on the road in lorries. Worst of all, this importation of Western materials and methods leads to centralization and urbanization. In order to benefit from these new developments, people have to crowd together in tightly-packed cities; suddenly, they are cut off from the natural environment; the land which gave them mud-bricks for free is now further removed, so they have to use and pay for a lorry to get them.

Even in narrow, Western economic terms the new ways are less economical than the old. But in a broader perspective their effects are disastrously expensive. Health decreases in the city, with a rising incidence of hepatitis and stomach problems hitherto unknown. The mental attitude of the people changes: alcoholism is another entirely new threat to health. Before people drank and got drunk at parties, in a happy social way. Now in the city there is a significant difference as people are trying to escape their daily existence and become dependent on alcohol.

Violence is on the increase. In more subtle but more basic ways the old interdependent society is attacked and the quality of life diminishes. The women who, in the traditional sector had a strong, almost equal position, suddenly find that the new industrialization has no place for them. The old are also excluded with no function and no room in the cramped space of the city. And the men now do the same monotonous work, at fixed hours, for eleven months of the year.

Along with industrialization comes the Western educational system, an integral part of the monoculture, but with no well defined origin; it comes with technology and a concomitant economic system. Children struggle with the Iliad, and don't learn how to make shoes from yak hair, or how to build an adobe house. If they learn how to build, it is as an engineer with concrete and steel. If they learn how to make shoes, it is from plastic in a factory. If they learn how to grow barley, it is out of books based on the monocultural system, with no allowance for local diversity. These books have no idea about the conditions at eleven thousand feet, and the wide variety of barley that has grown there, and all the local knowledge of minute differences in soil and climate which the local farmer is in tune with. The practical result is that the educated children cannot survive in the village. The only place they can live is in the city, as an urbanized consumer. If they have more education they have to go to Delhi, get more, and then they only survive in America or England.

Yet in Western terms, all this change is 'Progress'. All this economic activity increases the GNP, which in the traditional economy was virtually zero. The Western system is simply incapable of classifying traditional subsistence economies, and accounts them as worthless. Bhutan, for instance, which I visited in 1984, and which is very similar to Ladakh, was classified by the World Bank as among the poorest of the poor. And not only can modern methods of assessment not count the worth of traditional systems, they entirely discount the non-financial costs of the new—the disruption of people and the environment, the psychological effects of unemployment, the pollution of land, air and water.

. . . Conventional development is like a steam-roller covering the earth's surface with an artifical system—a way of life—which has no connection with the earth and its infinite variety or human cultures and their infinite variety. This progress is totally out of touch with that sacred principle, the interdependence of all life on earth.

However, there must be other ways of developing, where there is a real need for change; ways which do not lose sight of the fact that we are dependent on the earth. What I have tried to do in Ladakh is to show that there is an alternative: there is a way of improving the pre-industrial standard of living in such a way that it could become a model not only for developing areas but for us in the West.

In Ladakh, one of the very few real problems is the cold of winter. Traditionally, the winters are spent around a smoky dung-fire stove. Now, in the capital, those who can afford it have started buying imported coke to heat their homes. In order to pay for this, they have had to find a job in the modern cash economy, and forsake their work on the land. It is the beginning of a vicious inflationary spiral.

The first appropriate technology I helped to introduce was a system known as the Trombe wall solar space heating system, and it has been very successful. Even when the outside temperature falls to minus fifteen degrees centigrade, the inside of a Trombe room remains at about twenty degrees day and night, fluctuating only slightly. By now, there are about sixty Trombe walls in Ladakh; outside Leh, the Tibetan Children's Village has twenty houses and a hospital heated in this way, and there are others scattered throughout Ladakh. They are cheap—only the glass needs to be bought, and this can be done in Leh; the rest of the materials are the traditional mud-brick and wood. As a result, the walls blend very well with the traditional architecture and surroundings and more importantly, a farmer can raise his standard of living without losing his independence or disrupting his environment.

Excitingly, these initial efforts were greatly appreciated in Ladakh, and a local group formed around the project. The Group, known as the Ladakh Ecological Development Group, now has about a hundred members from all walks of life.

THE CENTRE FOR ECOLOGICAL DEVELOPMENT

We are very keen to ensure that we reach as wide a cross-section of people in Ladakh as possible, including those non-Ladakhis whose positions in the local administration give them a say in the region's development. It is not enough that we work only at the village level, at the grass roots. The political reality is that many of the decisions which affect village life are not taken in the villages themselves, but come out of committee rooms in Leh, or even Delhi.

The Centre for Ecological Development, which is the project's headquarters in the heart of Leh, serves the purpose of bringing our work, and the thinking behind it, to the attention of Ladakh's decision-makers, and thereby perhaps helps to promote the type of development strategy which we are seeking to encourage.

The Centre was inaugurated in 1984 by Indira Gandhi, shortly before her death. In the years since, it has generated very considerable interest not only within Ladakh itself but throughout the rest of the country. Members of the central Indian administration almost always include a tour of the Centre when they visit Ladakh, as do representatives of the media. In addition, the Centre has become something of a tourist attraction, with more than five thousand foreign visitors passing through its gates every year. His Holiness the Dalai Lama visited and consecrated the building in 1985, and we have also received both the Governor and the Chief Minister of the state of Jammu and Kashmir. Other visitors include ambassadors and senior diplomats from Delhi.

The building itself is an example of how traditional Ladakhi architecture can be 'updated' to meet changing expectations. Most of the building is solar heated. A small wind generator provides power for back-up lighting. In the garden, we have an array of solar cookers and dryers, and also a solar greenhouse. All these 'technologies' are in active use, and are accompanied by written and graphic material explaining how they work, and how they are built.

Inside the building, there are posters describing our work and information about specific issues in which we are involved. There is also an increasingly impressive library, which seeks to show the extent of the worldwide trend to more human-scale, ecological ways of living. We have books and magazines on organic agriculture, community schools, natural systems of health care, and self-reliance. We have critiques of development and industrialization, and manuals on a wide range of appropriate technologies. There are also directories of environmental organizations in India and abroad, and newsletters from ecological groups around the world. Here is powerful black-and-white evidence that the thinking which we espouse is gaining ever more widespread acceptance throughout the world at large.

The Centre also incorporates a small restaurant, in which a wide range of dishes are cooked in our own solar ovens. The restaurant is particularly popular with foreign tourists. Their presence at the Centre serves two very useful functions. Firstly, they help to provide financial support for our work, and secondly, they show the Ladakhis that this is something which they consider important and worthwhile. A respectable-looking, middle-aged Westerner who stands admiring a solar cooker, or spends an afternoon or more (as many tourists do) browsing through the library, is in fact making a valuable contribution to our cause, and is helping in the long run to promote more sustainable development strategies in the region.

We have recently completed some much-needed extensions to the building, so as to allow for increased office space and a meeting-room capable of holding up to one hundred and fifty people. In addition, we now have a basic engineering workshop (including one of the very few lathes in Ladakh), which enables our technical programme to operate more efficiently than was possible in the past. In the near future, we will be establishing a 'living museum' of traditional crafts, in which craftspeople from the villages will demonstrate such skills as weaving, stone carving and 'thangka' painting.

Much of our work at the Centre is undertaken jointly with the Ladakh Ecological Development Group. However, we make sure that the building is made available for as wide a range of local events as possible. We frequently sponsor meetings of the Cultural Forum, a group of local poets, writers and scholars concerned with the preservation of Ladakhi culture, and hold musical evenings at which Ladakhi musicians can play together with visiting Westerners. (Such informal events can serve the purpose of encouraging real 'cross-cultural' dialogue, and help the Ladakhis to get a better feel for life in the West.) In winter, the solar-heated library is a popular venue for all sorts of social and cultural activities, and thereby serves to promote the whole concept of appropriate technology.

Address by the President of Ladakh Ecological Development Group

TSEWANG RIGZIN

December 1986

'Tashi Delegs'

I, on behalf of the members of the Ladakh Ecological Development Group, offer my thanks to the chairman of the Right Livelihood Foundation for this Award to our group.

Our group was founded in Ladakh by Miss Helena Norberg-Hodge. During this short period of time we could do a little service to our people with the support of our well wishers throughout the world.

The main aims and objectives of the group are as follows:
1. To protect the precious traditional culture of Ladakh, which has been flourishing for centuries.
2. To demonstrate ways and means under which ecological balance can be maintained in the region.
3. To try for sustainable and longlasting development and maintaining tolerance among different faiths of the region.

To achieve the above aims and objects the group has started some activities such as publishing booklets and newsletters, arrangements of symposiums and seminars to educate the people, demonstration of solar ovens, greenhouses, windmills, etc.

It will, however, be difficult to get the people to follow the instructions of the group unless and until the members themselves follow practically. Our members are trying wholeheartedly to become practical in respect of our aims and objects.

For instance I am a farmer, mostly cultivating vegetables. I have started organic farming during the last three years. I have introduced new and different varieties of vegetables and herbs, which can grow successfully in Ladakh during summer and in greenhouses in winter.

The group hopes that the little services tendered by us will help Ladakh people in particular and the world at large in getting peace and prosperity.

Chipko—From Saving the Forests to the Reconstruction of Society
SUNDERLAL BAHUGUNA
December 1987

THE ILLITERATE WOMEN of the remote Mandal village in Uttar Pradesh, Himalaya, would never have dreamt that their innocent action to save a few ash trees in a neighbouring forest would become a source of inspiration to those all over the world who want to revive our planet. When the forest department refused to allot one ash tree to the villagers to make a yoke—an agricultural implement—and allotted fifty trees to a sports goods company, the villagers said: 'We won't allow them to fell our trees. If they come with axes to fell the trees, we will bear the axes on our bodies.' The situation did not arise, and the villagers never actually hugged trees. They demonstrated, and the forest department changed its plans and allotted trees in other forests where the villagers repeated the same words. They said: 'We will hug the trees.' Chipko is the Garhwali word for hugging, meaning 'to embrace'.

When, after the villagers' demonstration on April 23 1973, I asked a village woman, who had participated in it, as to how they had decided upon this action, she said: 'Look here, brother, imagine I am passing through a dense forest with my child and a tiger or bear attacks, what should I do to save the child from the attack of the beast? I shall hug the child, hold him fast and bear the attack myself.'

But this did not happen all of a sudden. There is a long story of India's cultural history behind it. Sages and seers—the spiritual and intellectual leaders of the society—lived in the woods in Ashrams with their disciples. They led a life of austerity and penance and pondered over the problems of humankind. Living in close contact with nature and in perfect harmony with their surroundings, which were forests, wild animals, rivers, and the mountains, they developed a philosophy of life in which they:

▷ saw life in all living beings and in all creation
▷ respected and saw divinity in all life
▷ regarded austerity as a symbol of bliss.

One of the sages experienced that a tree was equal to ten sons. He might have experienced that the ten important gifts of the tree were: life-giving oxygen, water, soil, food, fodder, medicine, fibre, and shade while the tree is living, and timber and firewood when it is dead. The stream of these ideas remained floating in the hearts of the people since time immemorial. Whenever they forgot this eternal truth, social revolutionaries like Gautam Buddhas, saints like Zambhoje in Rahasthan and Nund rishi (Sheikh Nur-ud-din-Wali) in Kashmir appeared on the scene and they reminded society of this rich heritage through their teachings in spoken languages. When India awoke after a long period of slumber during the British rule, the leaders of the society looked to their past for the solution of their problems. This gave birth to a renaissance in the middle of the last century. Raja Ram Mohan Roy was the pioneer of this renaissance and Mahatma Gandhi took it to its heights. There were great men like Dr Jagdish Chandra Bose, who proved the ancient scriptures' claim of life and feelings in trees and plants through his scientific experiments; and like Rabindranath Tagore the great poet, who as early as 1909 reminded Indian society and the world of the message of Aranya (forest) culture, when he wrote his famous essay *Tapovana*.

Conflicts over forest resources became a major socio-economic issue in India when the British rulers took over the best forests and 'reserved' them for commercial use. These forests were previously managed by the village communities. The people revolted against the government policy of usurping the forests, and against commercialization. These revolts were ruthlessly suppressed. In the native state of Tehri-Garhwal, the people of the Yamiena valley, inspired by the famous Dandi march of Mahatma Gandhi, came out in an open revolt and established their parallel government in a smaller part of the princely state. On May 30 1930, a big assembly of the people was being held on the banks of the river Yamuna at Tilari. They were fired on by the army of the ruler. Seventeen of them died and eighty were arrested. After independence, this spot became the place of pilgrimage for the hill people. They used to meet there to pay homage to the martyrs and discussed their forest problems. In 1968, they declared this day as forest day and approved a charter of forest rights, which in succeeding years was repeated in many villages. They pledged to revive the friendly relationship between the forests and the forest dwellers which was disturbed by the commercial management of the forests. They demanded an end to the contract system of forest exploitation and its replacement with local peoples' forest labour co-operatives; availability of raw material to local forest-based small industries; revision of forest settlement; and inclusion of forestry in the curriculum of education from the primary to the university stage. These ideas were taken from village to village by a band of Sarvodaya movement workers under the inspiration of India's walking saint, Acharya Vinoba Bhave, and the guidance of Gandhi's English disciple Sarah Behn, who were working in this region. Gandhi's other English disciple Mira Behn, the daughter of Admiral Slade, had many years ago, seeing floods in the Indo-Gangetic plain, warned about the disastrous effects of deforestation in Himalaya on the ecology and economy of the Indian sub-continent.

This was the background of the April 23 1973 event, when the women in Mandal won and the sports goods company had to quit. The message was taken to the whole region through the media of folksongs and religious discourses during foot marches for seven years from 1973 to 1980. There were confrontations between the villagers and the axemen. The government's plea was that felling of trees was done on the basis of scientific working plans. Chipko, by preventing the legitimate felling of trees, was doing a disservice to the country. The activists were branded as enemies of science, development and democracy for challenging the slogan of traditional forestry:

> *What do the forests bear?*
> *Resin, timber and foreign exchange,*

by expressing the scientific truth in its famous slogan:

What do the forests bear?
Soil, water and pure air.
Soil, water and pure air are the basis of life.

Thus, after the long sufferings of the hill women, this was recognized by the Indian Science Congress in its Varanasi Session in 1981 and thereafter by Mrs. Indira Gandhi (then Prime Minister of India), who ordered a ban on the felling of green trees in the Uttar Pradesh hills in an area of about 40,000 square kilometres.

This success of Chipko gave the activists a breathing time, and the message was taken from Kashmir, in the western Himalaya, to Kohmia in the eastern Himalaya through a 4,870 kilometre trans-Himalayan foot march in eight hill states of India, Nepal and Bhutan. This made Chipko well known in the India subcontinent. Like a migratory bird, it flew from the Himalaya to the southern state of Karnataka in September 1983. Now several Chipko actions in many parts of India—especially Gandhmardan, Orissa, Bastar, Madhy Pradesh, Aravalli, Rajasthan, Wayanadu, Kerala, Kodagu, Karnaraka, Kashmir and Himachal Pradesh are in progress. At some places the people are fighting a non-violent struggle to save their forests from mining, at others from being submerged under the dams, or from being converted into commercial plantations of eucalyptus and tea. In desert areas they are planting trees.

People's action has created consciousness in all sections of the society. There is a growing concern about the fast disappearing forests of India. The new forest policy of India is under review and the ecology aspect of forest management will be given top priority. Chipko's vision, in short, is:

1. Protect the remaining forests and declare water as the main product of the forests. The hill catchment areas of the river in Hamalay, Western Ghats and all other hilly regions should be declared as protected forests to keep the flow of rivers steady. There should be no commercial exploitation of these forests.
2. Monoculture plantations should be converted into mixed forests of indigenous species. Non-wood products of the forests,

especially products which give food, fodder, fertilizer and fibre without damaging the tree, should be encouraged.
3. There should be community control over the forests.

Although a massive programme of social forestry and wasteland development has been launched in India, forestry in India is still haunted by the two evil practices which are the legacy of colonial rule. These are: the raising of commercial plantations, especially pine, eucalyptus and poplars, which are not good for the soil and water; and policing the forests. The multi-million-dollar forestry projects of the World Bank, and of international funding agencies like the Swedish International Development Agency have been criticized on the grounds that, instead of growing fodder and fuel, which they professed, they planted species intended to produce raw material for industry. The slogan of the J&K Social Forestry Project of the World Bank was: 'Plant Trees and Grow Money'. In Karnataka, the small farmers went to the extent of uprooting ten million saplings of eucalyptus, because the farmers saw that the eucalyptus plantations by the neighbouring big farmers were adversely affecting their ragi (coarse grain) fields. These agencies and the governments have now realized their mistakes. When these issues were discussed recently with the Chairman of the World Bank in New Delhi, he said to the representatives of voluntary agencies, 'You become our eyes and ears.'

The Chipko Movement crosses national boundaries, and Chipko techniques to save forests are adopted by activist groups in many countries, in particular by 'Herbstwald' in Switzerland (1984), 'Save Danube Wet Forest' in Austria (November 1984), 'Earth First' USA (1988), 'Myers Island' Canada (1985), Germany (1984), and Sarawak in Malaysia (1987). The activist group in Göteberg, Sweden, who are protesting against the construction of an international highway through a dense forest in the West Coast, call themselves 'Tree huggers'. It has become synonymous with environmental protection campaigns.

Mr Rajiv Gandhi, Prime Minister of India, while addressing the United Nations General Assembly on Environment and Development, mentioning Chipko, said: 'There is the renowned Chipko

Movement in the Himalayas, where women prevent the wanton felling of trees by throwing themselves protectively around tree trunks.' A large number of messages expressing joy at Chipko being nominated for the Right Livelihood Award bear testimony to its popularity. The message from a veteran forester, S. Sohan Singha, retired Chief Conservator of forests and a teacher at the Forestry Institute whose students are holding the highest posts in the forestry departments of India, Nepal and Bhutan, reads: 'The next decade will find us revising our forestry textbooks and syllabi for forestry schools to accept and adopt Chipko's philosophy of forestry.'

Chipko does not confine itself to forests alone, but it challenges the present day development which is responsible for the triple threats of war, pollution and hunger to humankind.

The solution of present-day problems lies in the re-establishment of a harmonious relationship between man and nature. We shall have to digest the definition of real development to keep this relationship permanent. Development is synonymous with culture. When we sublimate nature in a way that we achieve peace, happiness, prosperity and ultimately fulfilment along with satisfying our basic needs, we march towards culture. We are not achieving the sublimation of Nature, but are butchering her in a quest to satisfy our never-ending desires. So we see perversion (*vikriti*). War, pollution and poverty are the forms of that perversion. War is the collective manifestation of individual dissatisfaction. Countries on the top of material prosperity are the biggest warmongers. They have diverted the dissatisfaction of their citizens towards war psychosis. If we want to replace war with peace, individuals will have to adopt austerity—the ending of desires. Individual satisfaction is then manifested towards social peace.

I want to make a clarification about basic needs. Once food, clothing and shelter were regarded as basic needs, but now oxygen has become our first priority. Our next demand is for clean and living water. We need fertile top soil to fulfil other basic needs.

In the present centralized system of production, it is ridiculous to think about the easy availability of these, but the truth is that no big stock of oxygen can be built. A person needs at least 16 kg. of oxygen in a day. Russian scientists have found that our oxygen requirements have become higher than the normal, because we now live in a technosphere, whose oxygen requirements are fifteen times more than all living beings. Is it possible to get so much oxygen free of cost in a market economy?

The same is true of water. We get our water supplies in big cities from reservoirs. According to the researches made by Austrian physicist Shallberger, the living element of water dies as soon as it is impounded or piped. We all are drinking dead water and even so its availability is decreasing. At the beginning of India's Sixth Five Year Plan there were 17,112 villages with water scarcity in Maharashtra. It is expected that 15,302 villages will be provided with drinking water during the Seventh Five Year Plan, but the underground water level is lowering fast in problem villages, and their number will be 23,000. There is water scarcity in the hill districts of Uttar Pradesh. Out of 2,700 drinking water schemes implemented by the State Government, 2,300 have become unsuccessful due to drying up of the sources. In West Bengal 2,300 persons died due to gastro-diarrhoea and water-borne diseases at the beginning of 1985.

The problem of soil erosion and salinity is also very serious. In India at least one-fourth of the total land area, fifty-eight million hectares of agricultural land and six million hectares of forest land, is seriously affected with acute soil erosion. The flood-affected areas have increased from twenty million hectares to forty million hectares during the decade from 1971 to 1980. There is danger of salinity to twenty million hectares and water-logging to ninety million hectares of land irrigated by canals. Whatever increased production has occurred is due to the use of chemical fertilizers and to irrigation. The chemical fertilizers, and mainly the nitrogen, have made soils addicts and the living soils as dust bowls. This system has made some countries very rich but their land has become very poor. When the demands of a growing civilization exceed the land's capacity to recover, soil erosion begins. Soil erosion is the symbol of the disparity between the society and its surroundings.

The way to overcome this crisis was shown by our culture. Nature is the source of permanent peace, happiness and the prosperity of humankind. Happiness and peace are related to living in harmony with nature. The first principle of this is that our living should be more and more dependent on renewable resources. These resources are derived from pastures, forests, croplands and oceans. The capacity of these to regenerate is declining due to over-exploitation, and production is going down. Whatever is shown as increased production was not a real profit but part of capital. The main problem today is to halt the decrease of capital by building up a lifestyle in which the harmonious relationship between human beings and nature is re-established. The milk of Mother Earth is not only for human beings, but for all her children—all living beings. It is why the poet has described the Earth as one who rears all. The whole Earth is a family.

A most practical way in this direction is tree farming. We will get rid of the problems of air and water pollution, erosion, salinity and water-logging in this way, besides getting enough oxygen and clean drinking water. The humanitarian scientists have arrived at the conclusion that the only solution to the problem of increasing population and decreasing cropland is to produce more on less land. When the land is used to produce animal-protein, we get one-hundred kilograms of beef from one acre of land in a year. The same land will produce one to one-and-a-half tons of cereal, seven tons of fruit, ten to fifteen tons of walnuts. If we can get trees giving edible seeds for humans and cattle, we can grow fifteen to twenty tons on the same area besides increasing the fertility of soil. Tree farming is the only way to heal up the wounds of Mother Earth created by the ploughshare during the last 10,000 years of agriculture. This will give an opportunity to human beings, animals and birds to survive together. Humans who are desperate to have the company of nature in the midst of affluence will not have to establish botanical gardens and zoos. The whole earth will become a garden—a garden where oxygen, water and food will be easily available to all living beings without any effort. This will certainly create a favourable environment for creativity—the development of art and literature.

The basis of Western civilization has been industrialization. Often the question is raised, what will take the place of industrialization?

I have already clarified that consumerism is anti-culture—perversion—because it makes human beings the butcher of nature.

There will be industries to fulfil the basic needs of human beings, but barring a few big industries which cannot be run on a small scale, all the industries will be decentralized. Every family and the village will become the temple of industry. This definitely needs a revolutionary change in technology. Instead of capital-intensive technology, labour-intensive pollution-free industry will have to be developed. The energy requirements of this will, as far as possible, be met with renewable resources—human and animal power, water, wind, solar and biogas, etc.—on a decentralized basis.

What Mahatma Gandhi dreamt of earlier this century, it seems, is the only alternative for the survival of humankind sitting over the store of arms. The two superpowers are negotiating and listening to the voice of non-alignment and disarmament, raised from Gandhi's country, to end the arms race. Though there has been no change in their activities, yet more optimistic is the voice raised in the affluent countries against their lifestyle. The most powerful voice is that of the Green Movement, which has made a place for itself in the politics of West Germany and is influencing national policies. I had an opportunity to participate in the First National Convention of German Environmentalists held in Wurzberg in June 1986. At the entrance of this conference was an exhibition on the life and work of Mahatma Gandhi. It seemed from their declaration as if the Hind Swaraj of Gandhi was speaking in the context of present-day global problems.

Gandhi had clearly said that he did not claim to give birth to new truths and that he only tried to apply the eternal truths in his own way to day-to-day problems.

Gandhi's speciality was to present alternatives. Authority may be replaced with service, wealth with austerity, arms with peace, and ideology with good behaviour. It is strange that Gandhi did not accept ideology, which has for centuries been supporting in one way or the other authority, wealth and arms. He presented good behaviour as an alternative to ideology. He emphatically said: 'My life is my message.' There can be no better presentation of

Indian culture than this. The new society will stand on these four pillars. In this way a new global view is emerging about which Dr Radhakrishnan had said: 'The future of humankind lies in finding co-relation with the past and future and with East and West, and establishing co-ordination between these.'

The practical way of achieving this co-ordination will be the coming together of humanitarian scientists, of social activists impatient for a change, and of compassionate literary men on a common platform. The wrong direction of science has given birth to the demons of war, pollution and poverty. So, there is need of such scientists to come forward, of people who instead of letting their knowledge be exploited for the selfish interest of those in authority, act according to the dictates of their souls for the welfare of all living beings. In the absence of their active support, the efforts of social activists impatient to bring changes are being weakened. We had such bitter experiences in the Chipko Movement for many years, when we declared soil, water and oxygen as the products of forests instead of timber, which is a dead product. We were branded as enemies of science, development and democracy. But our slogan, 'What do the forests bear? Soil, water and pure air', had come out of the wisdom of common people, and was decried till it was recognized by the Indian Science Congress in 1981.

The best type of art and literature has always been inspired by high idealism. What other great ideal in our era can there be than the establishment of harmonious relationships between Man and Nature, and getting rid of war, pollution and poverty. If this ideal catches the imagination of literary men, artists and journalists, they will move the hearts of masses for a march towards a new revolution. The crisis of civilization is a perversion born out of the rape of nature in the name of development; and the message of culture is the sublimation of nature to achieve peace, happiness and prosperity for all living beings.

This seems an uphill task, but it is possible. In our culture a noble mission is defined as 'yajna', and for the success of 'yajna' a co-ordination of knowledge, action and devotion is essential. I regard humanitarian scientists as the symbol of knowledge, social activists as the symbol of action, and literary men, artists and journalists touching the hearts of people as the symbol of devotion. They are in a minority, but this is a creative minority. Arnold Toynbee, the famous historian, in his concluding remarks on World History, said that the only hope for humankind standing on the brink of destruction is a creative minority. I invite all good-meaning and thoughtful people to join me in this minority for the restructuring of humanity.

Indian Woman and her Role in the Chipko Movement

INDU TIKEKAR

December 1987

ARE INDIAN WOMEN in any way different from their counterparts all the world over? Not apparently. Particularly as the victims of exploitation, dishonour and cruelty throughout human history, they are on a par with women everywhere. We find that in a society where physical strength, material wealth and sensual values took the upper hand in relationships, women are the main sufferers. India today is not exceptionally different from other countries where these conditions prevail. And yet, there is a qualitative difference in the Indian woman's reaction to situations in life; while facing challenging occasions, she shows a rare bravery, self-confidence, fearlessness, love and clarity of understanding.

What is the reason behind this uniqueness? Indian culture, since ancient times, has been a view that perceives divinity in every expression of life, and therefore has reverence for it. Woman is revered not merely as the symbol of creation and nourishment, ideally, she is wisdom herself. This does not mean that the fair sex was never treated secondarily. As a symbol of sensual pleasure and a thing to be owned, she was to be avoided by the virtuous and the seekers after truth. But this was a partial and dim picture of woman. In her fuller vision, woman always commanded honour. In such legends and religious scriptures as Durga or Kali the Goddess of Death and Destruction, woman alone could vanquish the demonic forces in the world. Uma, as the Goddess of Wisdom, revealed to gods that the spiritual oneness of life is the real strength behind their victories over evil forces. Laxmy, the spouse of the all-pervading God Vishnu, is the Goddess of Wealth and Plenty and stays in the untiring hands of those who work for the betterment of all.

Due to this high tradition, it was easier for Gandhi to give a call to the women of modern India to join the great struggle for independence. Not only prominent women like Sarojini Naidu and Kastuba Gandhi, but hundreds of ordinary women participated in the *satyagraha* movement—non-violent non-co-operation with the rulers to win India's freedom. While asking them to go for picketing at the liquor shops, Gandhiji used to emphasize the purity of woman's character that cleans the dirt of society. Disinterested concern for all, perseverance and capacity to sacrifice all with full abandon, were the qualities that inspired many during the freedom struggle. In all walks of life awakened Indian woman contributed her share in the renaissance of modern India.

It was because of this that free India never witnessed fighting for their rights in political and social life. Equality in all spheres is a natural growth in India. Even though backwardness and superstitious views are still being fought, 'feminism' as such is never going to root in Indian society to uphold women's cause. All 'isms' for that matter are only partial approaches towards the living truth of life, and if wholeness in every sphere of life is a sign of health, division of interests in the human or natural worlds is a sign of danger to peace and serenity. Man and woman are not—cannot be—opposed to each other, they together constitute humanity though their roles differ to a certain extent due to natural distinctness. Woman as mother of society should be above limited and narrow interests.

We find that in the new ecological movement of India's Chipko all these factors have played their part through the hill women in the Himalayas. Himalayan economy rests mainly on the responsible shoulders of women. For ages and ages they have suffered the hardships of uphill farming and cattle breeding. Housekeeping, serving the aged and the ill, bringing up of children, all these are, of course, their usual duties. They live and die for others. They have never spoken about their sacrifice, in fact, they know not the cultured terminology to express themselves before the so-called civilized world. They have not even heard of the comforts available to their counterparts in the big cities of affluent societies. Self-abnegation is so complete in their lives that they barely get any moment to think about themselves. But when the sheer necessities of life—water, fuel and fodder—became scarce in recent years and claimed all their energies, they became desperate. A few of them were driven to suicide, some lost their young ones, waiting and waiting for mothers in closed huts. And so, many spontaneously joined the Chipko movement and gave it a unique character.

Fifteen years earlier, when the Chipko-embrace-the-tree movement started, the social workers had some demands for the economic welfare of the people. They wanted to acquire the right to tap resin from Chir pine forests for the local small-scale factories. When women had to join the activists, it was a sudden decision. When menfolk were away, social activists had no information, and the labourers of the contractors entered the jungles with axes to fell the marked trees, women had to rush, carrying their children, to the spot and prevent the labourers from cutting the flora. They declared that the forests were their maternal homes, their abodes of rest and comfort.

The Gandhian *satyagraha*-spirit was already present in the activists. Women too picked up that spirit easily. When the contractors sought the help of police, they explained that they had no grudge against the police, nor were they enemies of the labourers. The real enemies are the wrong policies, the vested interests and the sensual values of life. Forests primarily bear water, soil and pure air. They provide food and fodder, fuel and fibre. Present-day consumerism has denuded our forests, it has impoverished our homes and fields.

Inspired by the centuries old example of the Amrita Devi and her followers in Rajasthan, who allowed themselves to be killed while trying to save their trees, men and women in the Himalaya, in their fight against the materialistic approach towards life, became prepared to suffer all hardships. They wanted to assert their right to survive through the survival of their ancestral homes. In the face of all sorts of provocation and encroachment, they never raised their hands in violence or abusive reaction. They have a faith that if one is ready to sacrifice all for a noble cause, one is bound to succeed.

These women had also participated in the prohibition movement in the Himalayas prior to the Chipko. Gandhiji's British disciple, Saralabahanji, was their inspirer and instructor. They had courted arrest and suffered long for the moral and economic welfare of society. They have shown that through their identification with noble and wider causes, instead of asking for equal status and fighting for their rights, they can win higher stature and command respect from all. Readiness to sacrifice, to suffer, demanding as little as possible, and honouring all with complete modesty means that one receives the fullness of life along with honour and satisfaction.

They have also suggested a new and alternative way of life through Chipko. Instead of running after more and more things to enjoy, if we limit our demands and cherish reverence for all life, nature will bless us through ecological health and there will be inner peace and poise. Human desires are never satiated through outer material gains, and if we wish to create a human world without exploitation, poverty and war, we have to stop not merely the race for armaments, but also to renounce exploitative, sensual ways of living. And in this, women can take a lead. As mothers, they know what it is to love unconditionally, how to give without asking, and how to create without destroying goodness. The boons of modern science and technology are today turned into banes through human selfishness. Will women folk come forward to embrace life on earth, reaching beyond the limiting identifications as male and female, and flowering into whole humanness? Men can follow their lead, and both can create a new world of mutual respect, real peace and beatitude. The Buddha, the Christ and Gandhi have already walked ahead. It is for us now to follow them. Let us take our inspiration from the Chipko women.

Food First: Beyond Charity, Towards Common Interests

FRANCES MOORE-LAPPÉ

December 1987

It was in 1974. 'World hunger' was splashed across the international marquee, making headlines, leading the evening news and sparking the now familiar, agonizing question: 'What can we do?' Famine in both the African Sahel and in Bangladesh became the immediate concern of the World Food Conference organized in November of that year. World leaders, government officials and corporate executives gathered in Rome, supposedly to draw up a blueprint for how to end hunger on earth.

By then, hunger had already gripped my attention for several years. *Diet for a Small Planet*—my first attempt to cut through the overwhelming complexity of world politics to show the needlessness of hunger—had come out three years earlier. So I thought that I, too, should attend the World Food Conference. I wanted to learn, to rub elbows with the real experts.

But something unanticipated happened. Flying back home that November, I knew that I would never be the same. The veil mystifying the 'experts' had lifted once and for all. I saw that the authorities, so schooled in the institutions of power, were unable even to ask the appropriate questions, much less point towards solutions. To them, hunger resulted from a simple equation out of balance—too little food for too many people. We needed only to correct the balance—more food and fewer births.

Yet in my own modest research I had already learned that more food could mean more hunger where people, deprived of land or jobs, have no claim to the increased production.

As I flew home from Rome, my self-doubts began to recede. I realized that if the 'experts' can't even ask the right questions, then leadership can come only from ordinary people—those able to look with fresh eyes precisely because we are not locked into institutions supporting the status quo. The implication hit me: If not people like me, if not me, then who can act freely to search for genuine solutions? And the answer came back, 'no one'. In other words, if the experts aren't 'taking care of business', then we'd better get on the job and fast!

I was then ready to take a bigger plunge. What made it possible was my extreme good fortune in meeting others who shared my sense of resolution. The following spring I met Joseph Collins, fresh from work on *Global Reach: the Power of the Multinational Corporation* and writing *World Hunger: Causes and Remedies*, an answer to the shortcomings of the Rome conference. Soon Joe and I were huddled in cramped offices over a supermarket just outside New York City. Writing furiously day and night, we were taking on the biggest challenge in our lives, but we didn't know where it would take us.

Our intense collaboration resulted in *Food First: Beyond the Myth of Scarcity*, a question-and-answer book taking on the most common misconceptions about hunger and its solutions.

When our publisher found out that our book was a serious treatment of the human-made roots of hunger, he demanded his 'advance' back. He had assumed we were writing a fad, 'how-to' book—how to slim down and save money during the imminent world food crisis! But he was too late—Joe and I had already invested our advance in establishing our new centre: the Institute for Food and Development Policy, which took its other name—Food First—from our book.

Our drive to launch this new education-for-action centre sprang from several observations. On the positive side, we saw hundreds of organizations popping up in the United States and in Europe—on university campuses, among the churches, in local communities—all focusing on the 'world hunger crisis'. We saw highly motivated, tireless activism. But because it was not backed by an analysis of the underlying causes of hunger, much of this energy was dissipated or even counterproductive.

Cutting through the simplistic and scary cliches about hunger, we had arrived at some surprising findings:

▷ No country in the world is a hopeless basket case. Even countries many people think of as impossibly overcrowded have the resources necessary for people to free themselves from hunger.
▷ Increasing a nation's food production may not help the hungry. Food production can increase while at the same time more people go hungry.
▷ Foreign aid often hurts rather than helps the hungry. But in a multitude of other ways we can help.
▷ The poor in the Third World are neither a burden to us nor a threat to our interests. Unlikely as it may seem, the interests of the vast majority of people in the industrial world have much in common with those of the hungry in the Third World.

We have sought, therefore, not just to change people's minds about particular, 'bad' policies. We've tried to affect the very unspoken assumptions we all carry with us about the way the world works and about our own interests in it. With food and hunger as our entry point, Food First's goal has been to change world views.

We work to provide a framework within which our individual acts take on greater meaning because we can see how they contribute to a larger movement for change. People often disparage their efforts as mere 'drops in the bucket'. Yet drops fill up a bucket quite fast—as long as there is a bucket! It is that bucket the Institute seeks to help construct—a more widely shared, evolving understanding of how we got to where we are and how together we can generate practical solutions reflecting our deepest values.

Thus, at Food First we have built four essential elements into our work, each unconvincing without the other: 1) a critical framework—an analysis of what's wrong, how we got here, and what is an appropriate response; 2) concrete lessons, *not models*, so that we can learn from the experiences of others; 3) specific guidelines to effective action, and 4) a clear articulation of the values guiding our work.

1. THE CRITICAL FRAMEWORK
First, we have sought to show that we humans can't blame nature for hunger. Not only has food production kept ahead of population

growth but the world faces price-depressing gluts of food. Still too often people believe hunger is caused when that simple equation is out of balance—too little food for too many people. But that view collapses once it's admitted that the greatest number of hungry people in the Third World live where food production has been increasing dramatically. In just two Green Revolution successes—India and Pakistan—more children died of hunger last year than in all Sub-Saharan Africa's forty-six countries combined.

We have sought to expose the real roots of hunger—not scarcity of food nor scarcity of land, but scarcity of democracy.... Democracy? How could we make such a claim, isn't much of today's hunger found in countries we call 'democratic'?

Yes. But in our view, no society has fulfilled its democratic promise if people go hungry, for by democratic we mean not only particular political structures but whether in the daily lives of its citizens key principles are manifest. These principles include: the accountability of leadership to all those who have to live with their decisions; and, second, the sharing of power so that no one is left utterly powerless.

Surely if some go without adequate food, they—by definition—have been deprived of all power, since eating and feeding our offspring come first for all living creatures. For these reasons, we say unequivocally: the existence of hunger belies the existence of full democracy.

We also avoid playing on guilt. We have found that guilt blocks people's innate compassion. While it can sometimes evoke quick results, it does nothing to change people's understanding of the world. Moreover, by dividing the world between 'we' who should give and 'they' who must receive, simple charity teaches us that our own interests diverge from those of the hungry—a perspective we believe to be false.

Since it can be shown that actual or potential production exists even where many people now go hungry, we must rethink the relationship of those living in the industrial world to those who now go hungry in the Third World. Such rethinking begins when we fully appreciate that, of course, people can and will feed themselves ... if they are allowed to do so. If they are not, we must assume that powerful obstacles stand in their way. The responsibility of those living in the industrial countries isn't to go in and 'set things straight' for them. It is to remove those obstacles. With US citizens, of course, we focus on those obstacles placed with our tax dollars—especially in the form of increasingly militarized US foreign aid. Today two-thirds of US foreign aid is military and security aid, often supporting the élite-controlled governments which are blocking the very changes needed to end hunger.

We first spelled out this message in our 1980 book *Aid as Obstacle: Twenty Questions about Our Foreign Aid and the Hungry*. It so shook up the US Agency for International Development and the World Bank, two of our primary targets, that each had to churn out internal responses to 'prep' their staffs on how to refute our findings.

Nowhere more dramatically than in Central America—with its obvious agricultural richness—can it be shown that hunger is caused by anti-democratic structures of power, not by scarcity. In the 1980s, the Reagan administration increased its economic and military aid to Central America sixfold. At the same time, we have intensified our campaign to show Americans that the conflict in Central America centres on people's 'right to eat', a right being denied by the very structures US aid is shoring up.

Just this fall—through the life and words of one courageous Honduran peasant woman—we have highlighted the immorality and futility of US aid to block change in that region. Our latest book, *Don't Be Afraid Gringo*, is the story of this Honduran peasant, Elvia Alvarado. Elvia has endured arrest and torture for organizing her neighbours to take land rightfully theirs under a 1970s agrarian reform law. With her book and a recent twenty-city tour, Elvia explained to Americans in all walks of life—even a group of US Army intelligence officers on their way to Central America!—the effects of US aid on her people. About US aid to her country, she told American audiences:

> We Hondurans are dying of hunger. We're not interested in fighting our neighbours. The national security they (our government leaders) are protecting is that of their own big stomachs.

They are protecting the fat cheques that come pouring in from the United States.

And to be released in January, our book *Betraying the National Interest* takes the next step. It shows how a militarized foreign aid policy predicated on fear of change in the Third World not only undercuts the interests of the majority who are poor there—and whose future depends on profound change—but works against the interests of the vast majority of Americans as well. We argue that the fate of the majority of people in the industrial countries is not in competition with the poor and hungry majorities in the Third World. Indeed, our security and prosperity depends upon their advancement.

There's nothing mystical about this insight. It is quite practical in today's interdependent world of globe-spanning corporate and financial institutions. The declining real earnings of US workers reflect their declining bargaining power as they are pitted against workers in the Third World who—denied rights to organize—must work for a fraction of what's considered a living wage here. Moreover, at least a third of US exports go to the Third World, but customers for the additional trade we so desperately need will be missing as long as the poor abroad lack the income even to buy what their own land produces, much less what our workers produce.

We explain to Americans a sad irony: their own tax dollars are going to shore up élite-controlled structures from Central America, to the Philippines, to Africa, that block changes which might allow an advancement for the majority, advancement so essential to our own security.

This fact of common interests we brought home very concretely in our path-breaking 1981 book *Circle of Poison*. It awakened millions of consumers to the hazards of pesticide 'dumping', disclosing that at least one quarter of pesticides exported from the United States were either banned, heavily restricted or had never been tested here. Used in the Third World on foods destined for import to the United States, the book exposes a 'circle of poison' threatening the health of both farmworkers and consumers. *Circle of Poison* helped spark an international network of activists, soon forming the Pesticide Action Network to halt pesticide dumping.

Of course, our most basic common interest is in peace. So we seek to show our fellow citizens that it's an illusion to imagine there can be a less volatile world without an end to hunger. It takes violence to keep people hungry. The poor do not go on passively watching loved ones die needlessly of hunger. They resist. 'I stand for peace,' a Central American peasant told me, 'but not peace with hunger.' Unless the universal right to the resources necessary to sustain life is acknowledged, hunger and violence will continue to mount worldwide.

People are horrified by the violence that often accompanies revolutionary change. But if we want to reduce the current violence of massive premature death resulting from hunger and the repression of those demanding their rights—as well as reduce the violence that revolutionary change can entail—there is one critical step that we as US citizens must take. We must stop our government's policy of strengthening élitist governments against their own people. This message is central to our work.

2. LESSONS, NOT MODELS

A critique of the economic and political roots of hunger and a framework with which to understand our common interests in change—both are necessary. But much more is needed. To overcome despair, we must ask: What can work? Where are people finding answers?

The temptation is to seek idealized models that can be transferred from one society to another. This is dangerous. When flaws come to light, despair is only reinforced. Besides, we suspect that economic and political structures are so culture specific that wholesale transfer could never work anyway. Instead of models we seek lessons—both positive and negative—from a wide variety of societies.

Our book *No Free Lunch: Food and Revolution in Cuba Today* is an example. The first balanced, in-depth study of the Cuban food system, it critically examines how this country has overcome hunger.

Working as part of the movement to halt US hostilities against Nicaragua, Food First does more than counter distortions by the US government; it analyzes in critical detail economic experiments underway in Nicaragua which might be useful lessons for any agrarian society involved in structural change.

While Washington portrays the Nicaraguan government as totalitarian, increasingly controlling the economy from the top, our book, *Nicaragua: What Difference Could a Revolution Make?*, reveals a highly pragmatic approach, encouraging private initiative. The keystone of the Nicaraguan agrarian reform was not to set a ceiling on the number of acres a farmer could own, as in so many reforms. Neither has state farm development been primary. Rather the guiding principle has been to attach an obligation to the right of farmland ownership. That obligation is to produce. 'Idle lands to working hands' was the early slogan of the reform. Its impact has been dramatic. Under the Somoza dictatorship, small farmers—the producers of basic foods—had access to only three per cent of the farmland; today they control more than forty per cent.

Because the 'Newly Industrializing Countries', such as Taiwan and South Korea, are so often held up as models of development, we're engaged in a major critical examination of these countries, asking: Are they successes—from the standpoint of the well-being of the majority—and are their lessons relevant to other poor countries?

But the lessons we are analysing and propagating aren't just those involving official government strategies. We are also working to contribute to the growing network sharing information about the most effective democratic development at the grassroots ... from the Working Women's Forum in India, to self-organized communal kitchens in Peru, to the 'green zones' producing food for the capital in Mozambique, to the union of sugar workers on the island of Negros in the Philippines. These and other examples are shared throughout our extensive network of colleagues worldwide and will be analysed in our forthcoming book, *Turning the Tables on Development*. We want to support those who are showing that 'development' isn't what happens when some outside aid agency enters the scene; development is what people themselves are able to achieve through democratic organizations, drawing on their own culture and ingenuity.

3. WHAT CAN WE DO?

At Food First, we believe education and action are not two separate kinds of work. We learn in order to act effectively; and in acting, we learn. We never describe ourselves as a 'think tank' but rather as a centre of 'education for action'. Our educational tools all urge appropriate channels for involvement. Our curricula for primary and secondary school students introduce even the youngest pupil to the idea of becoming an informed and inquisitive 'agent of change'.

Fortunately, the network of organizations to which we are allied is growing—more and more citizen organizations are taking up the many interrelated issues consistent with our framework of analysis. Millions of Americans are going beyond paternalistic charity in supporting efforts of the poor in the Third World. People who before might have been involved only in collecting money for relief are now setting up 'sister city' ties or establishing trade networks in which Third World producers retain a fair share of the proceeds from the sales of their products; or they are actively involved in human rights organizations bringing world attention to the repressive abuses those working for change too often suffer.

Our forthcoming book, *Making the Links*, highlights these and many other creative initiatives through which citizens in the industrial world are learning to see themselves not as benefactors but as allies of their counterparts in the Third World.

For those with specific questions, we've developed some specific guides. For the Americans who want to work directly in the Third World but not through government auspices, we have produced *Alternatives to the Peace Corps*. And for the many young people wanting to pursue an advanced degree but not within a stodgy graduate programme, training students to defend the status quo, we have *Education for Action: Graduate Studies with a Focus on Social Change*.

4. RECLAIMING, REFRAMING OUR VALUES

All of our work seeks directly to address people's feelings of powerlessness. Such feelings, we're convinced, derive not just from an objective assessment of the real forces weighted against change.

A lack of confidence can also stem from a shaky foundation—a weak grounding in clearly defined, compellingly expressed, and widely shared values. In the United States, we've been dismayed as progressive forces have watched on the sidelines while the Right has monopolized and manipulated a language and vision grounded in moral values.

With the election of Ronald Reagan and the rise of the Moral Majority, we began to ask how the Institute might begin to help progressives reclaim and reframe the very best in our culture's values as a stronger grounding for our work.

Of central importance is 'freedom'. For so long progressives have been accused of caring more about justice than freedom. Indeed, it's thought we would happily sacrifice freedom for greater justice! This is false. But to demonstrate why, we must define freedom. We must make clear that President Reagan's definition—unrestrained accumulation—is quite recent and contrary to much within our cultural heritage.

'What I want to see above all is that this country remains a country where someone can always get rich,' Ronald Reagan said in 1983. But in America's understanding of freedom, there's another, much older tradition to draw on. Many of our nation's founders defined freedom not as unlimited accumulation but as unfettered human development. And the two are incompatible. That's why Thomas Jefferson wrote in 1785 from poverty-wracked Europe, 'Because... inequality produces so much misery to the bulk of mankind, legislators cannot invent too many devices for subdividing property.'

From Jefferson to Franklin Delano Roosevelt to Martin Luther King, Jr., many great Americans have grasped that both freedom and democracy hinge on economic security. 'True individual freedom cannot exist without economic security and independence,' said Roosevelt. I am highlighting that understanding of freedom in my current project, a book-length dialogue about social values. To stimulate Americans to think through their own values, my voice confronts a mainstream conservative American who perceives security as the antithesis of freedom, and redistribution for equity as the seed of totalitarianism.

Using the classic form of a dialogue, I hope to demonstrate that a liberating social philosophy can stand up to the harshest conservative attack. Now being tested in classrooms, the draft manuscript is engaging university students in what I had hoped for most—intense debate about the values that guide our lives.

STRATEGY FOR CHANGE

Often we're asked why our offices aren't in Washington DC—the self-proclaimed centre of power of the Western world? And why don't we focus on lobbying?

Our answer is that until more Americans understand where their own interests lie and can see through their government's policies thwarting their own aspirations, no amount of lobbying by us could bring about the profound changes needed. So ours is a grassroots strategy. Believing that many well-meaning Americans want to be part of positive change, we seek to free them from beliefs that block action or lead to counterproductive work. Only then can they become the force needed to revive American democracy and move Washington in new directions.

We especially seek to reach people who reach other people—teachers, organizers, media spokespeople, church and civic leaders. They are the 'multipliers'—spreading our messages hundreds and thousands of times through their own work.

We do not seek to create a Food First movement per se. There are so many organizations and movements already working with their own communities on specific, concrete changes. We want to get better and better at serving those organizations and movements. We want to provide the tools—used by the classroom teacher, the consumer organizer, the journalist, or the church leader—that can reshape people's worldviews, enabling them better to effect change within their families, schools, workplaces, churches, citizen movements, political parties, and governments.

What keeps us going in our work? The courage of people we meet daily—yes, the Elvia Alvarados (whom I quoted earlier) but also the Midwestern Republican churchgoer who takes her first trip to Central America and from that point forward dares to be controversial.

What keeps us going? Witnessing people's lives change once food and hunger become their 'teacher'. We have seen students alter their career choices, adults change jobs, and people who had given up on being 'involved' who then discover meaningful involvement right in their own communities.

We have seen grade school teachers eager to teach about world hunger through our curricula because they help students see how they can be part of positive change. We have seen opinion-makers (even *New York Times* writers) shift away from the scarcity analysis of hunger, once exposed to our research and analysis.

We have seen food co-ops use our materials to incorporate far-reaching educational components into their services. We've seen church activists increasingly move beyond a plea for charity and reach out to their members with a message of our common interests with the hungry.

And most satisfying of all, we have seen Third World activists incorporating our research into organizing materials for their own grassroots campaigns for change.

The Food First message is not pleading, 'Oh, please come help share in the burden of saving the world!' Rather, we hope ours is a positive invitation—an invitation to participate in creating a framework for action that is visionary yet realistic. Recognizing the many formidable obstacles confronting us, Food First believes that by grounding action in our deepest values—freedom and democracy—human beings can build new forms of political and economic life more life-giving than any existing today.

And because we know that the Right Livelihood Award and all of those who have been honoured by it share this hope, I am proud to be here to celebrate our common work.

The Rainforest of Amazonia and the Global Climate

JOSÉ A. LUTZENBERGER

December 1988

I'M ACCEPTING THIS PRIZE not just in my name but for the Brazilian environmental movement, including those who are fighting for social justice, especially in the rainforest, now threatened with total obliteration within the next twenty years or so, unless present trends can somehow be reversed immediately.

We have no time, we must act now.

Among those who defend the tropical rainforest, many are still stressing the extremely important aspect of the loss of biological diversity. But this doesn't seem to impress and motivate those who have the power to act and to contribute to significant changes. To placate environmentalists, the Brazilian government is presenting them with maps that show dozens of little green dots on Amazonia, representing biological reserves or 'gene banks', as they also like to call them. Some researchers are helping them in this policy by experimenting with small islands of forest, left intact, but surrounded by seas of devastation, some of which are carried out by themselves for the sake of the experiment. They try to determine the minimum area of an intact ecosystem necessary for the survival of a given species. This sounds rational, but, if we accept all those little parks and reserves, and the philosophy that goes with it, then we accept that everything that is white on the map can be cleared away.

Of course parks and reserves, 'gene banks' and so on are necessary. We must protect what we can. Today, parks are often the only way of saving certain species or ecosystems. But to me the idea that we have to save parts of nature against our own destructiveness seems obscene. It is an avowal that something is profoundly wrong with our civilization. Shouldn't we also try to find out what is wrong with our present culture and how we can re-educate ourselves before it is too late? A healthy sustainable civilization can only be one that harmonises with and integrates into the totality of life, enhancing it, not demolishing it.

Modern industrial society has embarked on a course that, if allowed to continue much longer, will, in the end, destroy all higher forms of life on earth. One of the main aspects of how we wrongly deal with the world is reductionism, that is, facing only one issue at a time and thinking in straight lines. Looking for the minimum size of a certain ecosystem and then aiming at preserving only that minimum is a typical example. It completely leaves out the overall view of how those little green spots interact as parts of the whole, and what will happen once they are left alone in an ocean of devastation.

For the sake of this argument let's practise a little reductionism ourselves. We will leave out an important practical experience, namely that the few existing parks in Brazil today are all insufficiently protected. Most of them are being severely devastated. Quite recently great fires have almost wiped out two of our most important nature reserves. The government doesn't care. They keep saying there is no money for parks.

So let us suppose the new parks in Amazonia will be well protected, there will be no clearings, no hunting, no fishing, no fires. But when ninety-nine per cent of the rainforest will be gone, the rainforest climate will be gone. Our reserves will die.

And how will the disappearance of the world's rainforests, not only the tropical ones, affect world climate, especially in subtropical, temperate and subarctic regions, that is, mostly in the so-called First World?

Current discussions on and preoccupations with world climate are very much centred on the greenhouse effect and the ozone hole. But, apart from being an important sink for carbon dioxide when intact and a source of up to one thousand tons per hectare when it is cleared, the rainforest participates in other important mechanisms for climate control. Its fantastic evapotranspiration recycles rain water five to seven times as the prevailing winds move the clouds and masses of warm air from the Atlantic to the eastern slopes of the Andes, where they split, one half going north to northern Canada and into Europe, the other half south, to the southern tip of South America. Where the forest disappears and scrub or naked soil takes its place, evapotranspiration is replaced by hot updraughts that dissolve incoming clouds. If the present rate of devastation continues in the East, in the states of Para and Marahhao, it could soon trigger a collapse of the whole system, perhaps before thirty or forty per cent of the forest is cleared.

If the tropical rainforest disappears and the greenhouse effect makes the planet as a whole slightly warmer, it is, therefore, quite possible that northern and southern latitudes become colder, not warmer. Already we are having unusually cold winters in the south of Brazil.

Rather serious climatic changes could happen quite suddenly, not over decades and centuries. In the case of ozone, linear projections foresaw a slow and uniform depletion over several decades and all over the world. Instead, we now have the 'ozone hole' that comes and goes every year and becomes bigger and bigger. This kind of sudden effect we only get to know when it happens. Then it is too late.

What if the devastation of Amazonia triggers a sudden change in prevailing atmospheric circulation patterns? Or if the Gulf Stream that keeps Europe warm is somehow affected? We all remember the collapse of the Peruvian fishing industry when there was a sudden change in sea currents. Not only the fishing industry suffered, millions of sea birds starved to death.

The crisis had another sad consequence. It is partly responsible for the devastation of Amazonia. When Peru ceased to deliver fish meal for cattle feed—what lunacy, feeding fish to cattle!—a new market opened for soybeans. In the south of Brazil enormous soybean monocultures were promoted and subsidized. This was the end of our remaining subtropical rainforests and hundreds of thousands of small farmers and farm workers were uprooted. For them and people uprooted elsewhere, by the alcohol for fuel programme and by the big landlords in central Brazil and the North East, the 'Polonoroeste Project', with World Bank money, (that is, with taxpayers' money from the First World), opened western Amazonia for settlement. This triggered one of the worst forms of devastation in Amazonia, a devastation that now continues and spreads out on its own impetus and is almost impossible to stop.

Fortunately, even governments are now becoming aware and preoccupied with the danger of serious climatic change, as became clear in the two world climate conferences: in Toronto, 'The Changing Atmosphere', June 27-30, 1988, and in Hamburg, 'Climate and Development', November 7-10, 1988. But real danger is still seen as lying some fifty years in the future. One of the presentations in Hamburg even dealt with how agriculture should prepare for the coming changes, instead of proposing changes in agriculture now, in order to diminish the effects of modern agricultural methods on the factors controlling world climate. But most scientists at these conferences were in agreement that the climatic irregularities we have been experiencing all over the world are the beginning of the greenhouse effect.

Perhaps we should be rather surprised that things are not worse. Considering how modern industrial society massively interferes with the mechanisms of climate control—carbon dioxide, ozone, evapotranspiration, albedo, aerosols and dusts in the air—it is amazing how resilient the atmosphere still is. But, as we saw, sudden changes can happen anytime.

But we need no cataclysmic changes, such as a new ice age or the melting of the polar caps and rise of the oceans, to be in serious trouble. The continuation and aggravation of what we already see, (for instance, several summers in succession like the last one in North America), would be catastrophic. Humanity cannot afford a situation where we have no secure harvests anymore, even if the weather is very nice for the beaches!

Since arguments in terms of the beauty and diversity of the great symphony of life (of which we humans are only a small part) do not seem to impress the powerful, perhaps the impending climatic perils that were never as visible as they are now will make them act. The devastation, for whatever reason, of the world's tropical rainforests is totally irreversible. We will not be able to remedy the unpleasant consequences, but we might still be able to prevent the continuation of the devastation.

We can always learn from our mistakes. But do we have a right to risk mistakes that have unacceptable and irreversible results?

Africa's Food Security: the Preservation of Genetic Diversity in Crop Plants in Ethiopia

MELAKU WOREDE

December 1989

IT IS A GREAT PLEASURE INDEED to be honoured today with the Right Livelihood Award which I accept with deep gratitude and many thanks.

Although I started to actively work on genetic resources some twenty-three years ago, the motivation to do so goes back to my Freshman year in college, some thirty-two years ago. It started when a visiting professor from Oklahoma University in the United States delivered a speech on agriculture. I asked him why the big, well-developed countries were not giving us superior varieties of crops so that we produce them here? He answered by telling a story of a crew that was sailing on a sea, out of water supply. In desperate need of water, the crew kept on calling for help with the radio. Being advised to drop the bucket right where they were, the crew was surprised to know what they were sailing on was fresh water. And the answer given to my question was, 'Drop your bucket right where you are.'

I always kept this important advice in mind in subsequent years as I conducted research or taught agriculture, especially genetics and plant breeding, and in developing the genetic resource activities at the Genebank since I became its director in 1979.

The Ethiopian region is characterized by a wide range of agro-climatic conditions which account for the enormous diversity of biological resources that exist in it. Probably the most important of these resources is the immense genetic diversity of the various crop plants grown in the country. Not only does Ethiopia possess important diversity in crops domesticated elsewhere, such as wheat, barley, grain, legumes and several oil plants, it also has developed its own indigenous cultigens such as teff, sorghum, noug (*Guizotia abyssinica*), *Brassica carinata* and coffee, many of which are now of great importance.

The existence of such genetic diversity in Ethiopia has great significance for the long-term food security of the country, and for the rest of the world because it provides the resource base on which sustained development of high yielding and stable varieties depends. At present, the existing broad range of genetic diversity, particularly that of primitive and wild gene pools, is being rapidly depleted, displaced or abandoned due to causes that are many and complicated. Various factors have interplayed in posing such a threat which is progressing at an alarming rate.

The drought that prevailed in the last few years has directly or indirectly caused considerable genetic erosion, and at times has even resulted in the wholesale destruction of genetic resources. The famine that persisted over the years in some parts of Ethiopia has forced the farmer to eat his own seed in order to survive or to sell his seed as a food commodity, thereby posing the threat of massive displacement of native seed stock by exotic seeds introduced in the form of food grains donated through relief agencies.

In the midst of drought in the late seventies, the Centre began systematic scientific collection of threatened crop species. The genebank has now gathered some 48,000 accessions of a broad range of indigenous food, feed and fibre crops and plants of medicinal and industrial importance, supported by the Federal German Republic through a bilateral technical and economic development programme. These are all maintained following internationally established methods and procedures. Also because of such crises in the country, the Centre, in collaboration with the Ethiopian Seed Corporation (ESC), and with farmers and agricultural agents, has, since 1985, launched a seed reserve programme to cover four principal areas subject to recurring drought. Similarly, the Centre in collaboration with the Ministry of Agriculture has been undertaking germplasm rescue operations in nearly all crop growing regions of the country since 1987.

In Ethiopia, as in many other developing countries, farmers play a central role in the conservation of germplasm as they hold a major portion of the existing genetic resources. Peasant farmers always retain some seed stock for security unless circumstances dictate otherwise. With such traditional advantages at hand, the Ethiopian genebank in collaboration with farmers, agricultural extension agents, breeders and the country's seed corporation, is working to develop a farmer-based native seed conservation and enhancement programme which has been in progress since 1988. This was designed mainly with a view to save major cultivars from becoming extinct, as well as to help the Ethiopian farmer hang on to his crop diversity while improving his evaluation of each season's harvest: together, we expect to increase annual yields by as much as five per cent each year.

The cash Award from the Right Livelihood Foundation will go towards this work in Ethiopia, through the Unitarian Service Committee of Canada (USC/Canada) which is currently financially supporting farm-based conservation and enhancement work in Ethiopia and other developing countries through its Seeds of Survival Programme. The increasing realization by the international community that plant germplasm is the most important source of variation for any crop improvement, and that this irreplaceable resource is constantly threatened by extinction, has resulted in a corresponding global interest in its preservation and effective exploitation.

Many gaps still exist in the knowledge of and scientific approaches to conservation practices, despite the many recent advances in this area. An opportunity exists, therefore, to create new strategies and approaches to tackle the enormous qualitative and quantitative dimensions of those conservation problems unique to each country. Genebanks in developing countries should continue to make full use of such opportunity and seek to apply the progressively advancing technologies that the international community is providing.

Africa's food security depends upon a new approach to scientific research that unites farmers and scientists, governmental agencies and non-governmental agencies in a co-operative effort. Africa and the Third World have been, are and can continue to be the world's bread basket and not its basket case. The conservation and development of their genetic resources are, therefore, an important cornerstone to agricultural development in these regions and the world at large. As is already happening in my country, farmers and national genebanks in developing countries can work together to preserve and expand crop genetic diversity on behalf of all humanity.

Plant genetic resources are seldom 'raw materials'; they are the expression of the current wisdom of farmers who have played a highly significant role in the building up of the world's genetic resource base. The world cannot therefore ignore or deny the intellectual contribution of farmers, i.e. Farmers Rights—Informal Innovation System—must be recognized.

The Plant Genetic Resources Centre in Ethiopia has been a leader in Africa in drawing continental attention to the problems of genetic erosion through a number of training workshops, providing on-site consultation and international symposia which it has been hosting in co-operation with (and with financial assistance provided by) various NGOs, and especially the Rural Service Committee of Canada (USC/Canada).

As a further development of the African Committee for Plant Genetic Resources, of which I was the first Chairman, we are currently also creating a new African Commission on Biological Diversity as a tripartite organization including governments, scientists and NGOs in Africa. At the international level, we have been active in promoting the establishment of the Food and Agriculture Organization/Commission and Undertaking on Plant Genetic Resources (FAO/CPGR), actively working toward promoting Farmers Rights—the Informal Innovation System.

In Ethiopia genetic resource activity already represents a major national effort which the country has systematically undertaken over the last decade or so. The existing options pose a serious challenge, requiring major inputs in terms of technical know-how and material considering the country's rich and diverse biological resources. There is also a unique opportunity to salvage and effectively utilize crops which the farming community in my country has developed and maintained and which at present provide a major portion of the existing crop genetic resources.

The Award bestowed upon me is a tremendous support and encouragement to my country and myself in the effort to salvage these dwindling resources with a view to providing, in sustainable ways, useful germplasm to breeding programmes both in Ethiopia and the world community at large. It is also a tremendous encouragement to the staff of the Genebank and other relevant institutions and organizations who in no less small measure have contributed towards the success for which I am now honoured.

3

The Rights of People

INTRODUCTION TO THE PROJECTS

Evaristo Nugkuag Ikanan (AIDESEP)
1986
'for organizing to protect the rights of the Indians of the Amazon Basin'

Since the Spanish invasion of South America in the sixteenth century its indigenous Indian peoples have been murdered and all their rights violated by European adventurers, traders and soldiers, and their culture undermined by missionaries. The twentieth century has been no exception: since 1900 the Indian population of Brazil has fallen from one million to only 200,000.

EVARISTO NUGKUAG IKANAN, an Aguarunan Indian, is a Peruvian Indian leader who has dedicated his life to organizing indigenous people in the Amazon basin to secure their human, civil, economic and political rights. In 1977 he was one of the founders of the Aguaruna and Huambisa Council (CAH) in his home area, which, with a formidable and comprehensive programme encompassing health, economic development, education and legal defence, soon became one of the most successful indigenous organizations in Peru.

In 1981 Nugkuag was again instrumental in setting up, and became the first President of, AIDESEP (Inter-Indian Association of the Peruvian Rainforest), which brought together thirteen different jungle groups representing over half the 220,000 rainforest Indians of Peru. AIDESEP's programme covered the same issues as that of CAH and is also concerned with the defence of native lands and resources, social communications and publications, and general advice and assistance.

Evaristo Nugkuag Ikanan
AIDESEP
Av. San Eugenio 981
Urb. Sta. Catarina
Lima 13 Peru

Sahabat Alam Malaysia, Sarawak
1988
'for their exemplary struggle to save the tropical forests of Sarawak'

The Sarawak Office of Sahabat Alam Malaysia (SAM—the Friends of the Earth organization in Malaysia) has been caught up since 1986 with the native Penan people of Sarawak in a desperate struggle against logging in the province. In 1983 this logging was proceeding at the rate of seventy-five acres per hour, enabling Sarawak to provide thirty-nine per cent of Malaysia's tropical log exports, which amounts to over fifty per cent of the world's total. The logging is systematically destroying the culture and livelihood of the area's native peoples, including the Pelabit, Kayan and Penan peoples. In 1987 the forest people called for the recognition of their rights to the land and for the cessation of large-scale logging. In a series of non-violent actions during 1987 and 1988 native people, especially the forest-nomadic Penan, succeeded in stopping logging activities in the major timber districts of Sarawak.

SAM Sarawak
PO Box 216
98058 Marudi, Baram
Sarawak, Malaysia

SAM
43 Salween Road
10050 Penang
Malaysia

Inge Kemp Genefke
International Rehabilitation & Research Centre for Torture Victims
1988
HONORARY AWARD
'for helping those whose lives have been shattered by torture to regain their health and personality'

In response to Amnesty International's appeal in 1973 to the medical profession to help fight torture, Dr Inge Kemp Genefke formed the first Amnesty International medical group in Denmark. Its pioneering investigations into torture and its consequences for its victims led to the establishment of more medical groups, and by 1982 there were twenty-nine such groups with over 4,000 doctor-members. The need for treatment and rehabilitation of torture victims then led in 1982 to the establishment of the International Rehabilitation and Research Centre for Torture Victims (RCT), with Dr Genefke as Medical Director, in Copenhagen.

Inge Kemp Genefke
RCT
Juliane Maries Vej 34
DK-2100 Copenhagen O
Denmark

Survival International
1989

'for the broadest, longest-serving and most effective campaigning and educating organization working with tribal peoples to secure their rights, livelihood and self-determination'

Survival International was founded in 1969 on the belief that tribal peoples are too often treated as obstacles to progress, objects of study, exotic tourist attractions, or potential converts, when in fact they are members of complex and viable societies with a sense of purpose, fulfilment and community that many in our 'modern' societies might envy.

There are some 200 million tribal peoples, just four per cent of the world's population, and, as has often been the case, many find themselves in conflict with dominant cultures. Survival International has three broad objectives:

▷ To help tribal peoples to exercise their right to survival and self-determination;
▷ To ensure that the interests of tribal peoples are properly presented in all decisions affecting their future;
▷ To secure for tribal peoples the ownership and use of adequate land and other resources and seek recognition of their rights over traditional land.

Survival International is organized into an International Secretariat, registered as a charity and based in London, and has consultative status as a non-government organization with the UN and EEC. It has national sections in France, Spain and the US and some 5,000 members in sixty countries.

Survival International works through projects and campaigns, education and publications. In 1988-89 Survival International worked on fifty-two cases of violations of tribal peoples' rights and seven field projects in Australia, Bangladesh, Brazil, Canada, Chile, Colombia, Ecuador, Ethiopia, Guatemala, Guyana, India, Indonesia, Kenya, Peru, Sarawak, Somalia, South Africa, Sri Lanka, Thailand, the United States and Venezuela. Major campaigns of the past have included:

▷ Pressure on the World Bank which has resulted in that organization giving considerably greater attention (if not yet enough) to the impact of its lending on tribal peoples;
▷ The 'Nothing to Celebrate in 88' campaign with aboriginal peoples in Australia to coincide with the bicentenary of European colonization;
▷ An International Day of Action for the Yanomami Indians of Brazil;
▷ A signature and public education campaign to stop the logging of tribal lands in Sarawak.

Stephen Corry, Director General
Survival International
310 Edgware Road
London W2 1DY
Tel: 071 723 5535

The Tribespeople of the Amazon
EVARISTO NUGKUAG IKANAN
December 1986

I ACCEPT THIS PRIZE on behalf of my people, the Aguaruna, and our organization, the Aguaruna and Huambisa Council; on behalf of the indigenous people of the Peruvian rainforest who together make up the thirteen organizations, councils and federations grouped within AIDESEP (Inter-Indian Association of the Peruvian Rainforest); and finally on behalf of all the tribespeople of the Amazon who are members of the indigenous organizations of Colombia, Ecuador, Peru, Bolivia and Brazil, who joined together in 1984 to form the Coordinadora of Indigenous Organizations of the Amazon Basin.

It is something very new, very special that the panel of this institution should have considered myself, a representative of the Aguaruna Indians, as worthy of such an important prize. It is new for two reasons:

In the first place, tribes peoples, such as myself, are not used to receiving recognition at an individual level. The great difference between the indigenous and the Western worlds is that we live a collective and communal life. We live in communities, we feel at home with our brothers of our group, of our people, of our indigenous nation. Together we are strong. The individual is important as a member of the whole. Our ancestors had no experience of Western individualism. We cannot conceive of it for our children. The confrontation between the two worlds, the Western and the Indian world, has had a great impact because we were not prepared for individualism, egotism and its most dangerous expression, capitalism, which is destroying a great deal of the world.

In the second place, this is the first time that an institution of the Western world has considered us for such an important and esteemed prize. Of course, some doors have been opened to us to voice our opinion. But this is the first time that our struggle has been recognized as legitimate, the struggle of all indigenous organizations who defend our ancestral claims to our lands and our right to be the bearers and creators of our own culture.

For that reason, the prize is a new development. In the West you have tended to glorify the great deeds of the settlers and conquistadores, of the pioneers who won hundreds of people over to civilization. How many monuments to Colombus and other heroes are to be found in the cities of Europe? We have been the great forgotten ones. Nobody has heard of the history, past and present, of the colonized, the history of the native people subjected to civilization.

Much has been said about the independence of America, that great step forward in the history of our peoples. But Bolivar brought his ideas from Europe and the independence was for those of mixed race. For us, the tribal peoples of America, the original Americans, the matter is still not resolved.

So far, not even in the United Nations, even less in the legislature of our own countries, have the collective rights of the people of our nations, which exist within those states, been recognized. That is why this prize, awarded by a Western institution, is doubly important. We are hopeful that it will be a step towards brotherhood while there is still time.

The arrival of so-called civilization meant the beginning of a series of transformations in the lives of our peoples, destroying, partially or totally, our lands and natural resources, imposing on us customs and cultural values alien to our own culture which allows us to respond efficiently to the problems and uncertainties of our environment.

Throughout this process of conquest and colonization, which began 400 years ago, people influenced by different ideas have left their mark of destruction on our peoples. There was the pioneering conquistadore in search of 'El Dorado', who saw in our forebears no more than obstacles whose elimination was a heroic achievement. Then there were the religious missionaries, who in the process of supposedly saving our souls, destroyed our spirit, leaving the door wide open to exploitation. And finally, the rubber merchants, who left it to their agents to physically exploit the Indian people to the limit of their endurance. The colonizers showed the world an empty Amazonia and showed more respect to the animal world than to the lives of our ancestors.

Our people were weakened, subjected to slavery, to alien diseases which over four centuries decimated our population. The atrocities committed on the Indians and their systematic extermination were hailed as triumphs of civilization. Of the seventy-eight nations or ethnic groups which existed at the beginning of this century in the Peruvian Amazon, no more than sixty still survive. Of the 45,000 people who inhabited the Huitotos territories at that time, today only 5,000, including myself, can be proud to receive this prize.

As a consequence of colonization, the Indian territories grew smaller every day and the natural environment which sustained our lives was irreversibly destroyed. When the governments recognized our communal titles, we Indians understood that the rest of our lands were no longer our own and that the long struggle for their recovery was only just beginning. Communal lands were divided up. People whose relationships formed the basis for their survival were separated. And governments turned a blind eye to the massive destruction of the very resources which through the wisdom of our ancestors had sustained us through the centuries.

In addition to this destruction, we Indians have been victims of the Western cultural superiority complex. Alien values, beliefs and needs were imposed on us and used as instruments of oppression. Those values are useless in our Amazonian environment.

Despite the destruction, we are still alive, although our culture has been weakened. But some of our people have reached such a state of alienation that they have become ashamed of their own culture. They have become dependent on the very system that oppresses them.

Many of our children have forgotten their native tongue and are experiencing a sense of shame and prejudice towards our traditional culture. The civilizing mission schools have gradually discredited the knowledge of our forefathers. The introduction of the employment system has disrupted our collective way of living. In a racist society, we are at the very bottom of the ladder.

Now, it may seem to you that all this should be past history and behind us now. But it is a history characterized by ignorance and oppression by a dehumanized Western capitalist system. This terrible history is the history of today and the missionaries and conquistadores, the big commercial companies and the tourists, are today's actors and live amongst us. I shall remind you of a few cases, and I am only quoting those which have concerned our organization, AIDESEP, over the last few months:

▷ Slaughter of Ashaninka Indian leaders in the Gran Pajonal region and wholesale theft of their lands by big business.

▷ Murder of Chief Nuncanquit of the Tsuntsuntsa community, one of my own people, the Aguaruna, who was killed in an ambush by gunmen hired by colonists who had been evicted by court order after a long and extremely costly trial.

▷ Displacement of the Ashaninka Indian population in the Atalaya region and occupation of their houses, leaving the Indians to their fate.

▷ Large scale special colonization projects in Pichis-Palcazu, Jaen, San Ignacio, Bagua, Huallaga Central, Alto Mayo and Madro de Dios, Aguaruna, Amuesha, Ashaninka, Amarakaere, Cocamilla and among other Indian peoples and which are being financed by Western institutions such as the World Bank....

▷ Ethnocide of the Nahua people in La Convencion who are dying of diseases brought in by the timber merchants as a result of which the numbers of Nahuas have substantially decreased.

▷ Large scale invasion of the lands of the Amarakaere by gold-mining companies who defend their plunder by trickery, bribery and arms sales....

I am speaking about Peru, about our recent problems. Our sister organizations in the Amazon Basin have similar problems, which are in many cases even more serious:

▷ Invasion by gold prospectors and mining companies in the Upper Rio Negro region in Brazil, endangering the survival of over 18,000 Indians.

▷ Oil exploration in the tribal area of the Javari Valley by the State company Petrobras, which is causing violent conflicts between the Indians and the intruders.
▷ Violent expulsion by the military police of the Pataxo HaHaHa in South Bahia, Brazil, whose long struggle has been frustrated by the neglect of the Federal Supreme Court.
▷ Ethnocide of nineteen Indian groups of approximately 10,000 people by the largest development project in the world, the Grande Carajas in Brazil which is financed by German and Japanese banks and the World Bank. . . .

I do not wish to tire you; the list is endless. Crimes, atrocities and injustices are too numerous to any longer affect our feelings. Can you imagine the outrage which any of these cases would cause if they were to happen in your own country?

And yet, our sons and daughters are children, just like yours, our women and our men are people, just like any other, our nations have their pride, their history, their heroes, their beliefs, their customs, just like any other. Today you are awarding us a prize because you have understood. So you can also stop your governments helping to plunge our people into further suffering. We shall grow stronger in our struggle with your help.

I would like to explain something to you. That huge green sea of Amazonia is not a paradise of fertility. We Indians live in an environment which for centuries has retained its ecological balance because we have applied our knowledge wisely so as not to destroy the land.

The lands of the Amazon are the poorest in the world and demand a lot of care if they are to continue to be fertile. Once destroyed, they will not recover. Only two per cent of the soil is suitable for agriculture and we, the Indians, have lived on those lands for centuries.

This is the reality which the various governments of the Amazon do not want to understand. They use the rainforest to avoid agrarian reforms benefiting the peasants, those who work the land in other areas of our countries. Through colonization, they are seeking to avoid a fair distribution of land for them.

Neither do the financial institutions of the developed countries understand, neither the World Bank nor the Interamerican Bank, the AID or any others. Many of their projects fail, because any changes to the rainforest in terms of agricultural exploitation will result in the complete destruction of the Amazonian land.

For capitalism, this is merely a failure, a bad investment which in the short-term has some political advantages. For us, who are thinking about the future of our children, it means the loss of our way of life. That is why our points of view differ. The settlers, the government, the banks, regard the Amazon as a means to make a quick profit, without giving a thought to the long-term implication of their decision. To us, the earth is the basis of our existence and we need to retain her whole with all the variety of nature and we cannot negotiate her price or forget about her. To us, the earth is not just a factor of production. But this is the way that the governments operate. They give us small communal settlements, dotted around colonization projects, surrounded by enterprises engaged in plundering our earth.

When the old Chief Juaneco of the Ashaninkas asked recently for our support in the titling of his community Tahuanti, we explained to him the difficulties of obtaining large extensions of land. He answered: 'You say this because you know us, this is our land. Do only people live here? No, there are also monkeys, even bears. And where would the monkeys go if we didn't ask for land for them as well? That is the way our land is. The land is for everyone: men, animals and plants, land full of the spirits of our forefathers, it is a reciprocal relationship. The land is for our men of today and for our children.'

How can governments allocate land to us when they arrived here after we did? That is why the struggle for land is the greatest struggle today. A struggle we cannot afford to lose because our life depends on its outcome. Why do I want you to know this? Because in that great green sea the Indians are beginning to learn the meaning of hunger.

And the other great battle is for respect. Because these two things go together. If we have our lands taken from us it is because we are seen as being different, we are not respected as equals.

What do governments, religious institutions, officials or the military do to defend our rights? Nothing, because they have other priorities. And that is why we, the indigenous people, have decided to give the process of unification priority. And we are getting stronger, from community to community, village to village, from country to country. The voice of the indigenous people is becoming louder.

Now I will present to you the indigenous organizations which in my name have been recognized under this banner:

1. THE COUNCIL OF AGUARUNA AND HUAMBISA, the indigenous organization of my place of origin which has united the work of eighty-nine communities settled along five rivers and tributaries of Maranon, near the border with Ecuador.

This council has joined together our two ethnic groups to mark a new beginning. Together we are defending our lands and resources. There are many cases, such as our first fight against the film production company of the German Werner Herzog, and then the American goldminer Gerald Stucky, who tried to plunder our resources and misuse our culture.

The Council is dealing with our health problems, which have been ignored by the state, with a programme of traditional medical cures involving more than 100 paramedics, obstetricians, supervisors and laboratory assistants. All these people are Aguaruna or Huambisa. There are more than eighty health clinics, five central clinics, there is a permanent training programme and an independent medical supply system.

The Council has organized an efficient economic system with more than seventy communal enterprises and five centres of commerce, as well as a mechanism for controlling surplus production which allows us not to take more than we need from our land.

Through the Council we are trying to give our children a meaningful education which relates to our culture, of which we were robbed by the Jesuit missions, the merchants and the Sumner Institute of Linguistics. Finally, through the Council we have recovered our pride in being Aguaruna and Huambisa.

In the present municipal elections, we have won all the mayorships in the region. Certainly now the government had to recognize us as the legitimate spokespeople for our region.

2. THE INTER-ETHNIC ASSOCIATION FOR THE DEVELOPMENT OF THE PERUVIAN AMAZON (AIDESEP) is also included in this prize. This organization unites the great majority of the organized bodies of the Indian population in Peru. It was created in 1980 through the initiatives of a group of federations, councils and indigenous organizations. Our principal objective is to promote an independent response to the day to day problems the indigenous population of Peru face and defend our rights from our indigenous perspective without outside interference.

Our principal line of action is the defence of our land and our resources as well as the right to our own language, culture and education, for self-determination and the right to political representation and for the security of our people.

We have developed indigenous policies for all those issues and are working towards a proposal for the Amazon which allows the establishment of alternative land use. In general, we are trying to give support to indigenous organizations *vis-à-vis* their dealings with governments and their Amazonian development agencies.

3. Lastly, the COORDINADORA, THE CO-ORDINATING BODY OF THE INDIGENOUS ORGANIZATIONS OF THE AMAZON BASIN, which was formed in 1984 by the most representative indigenous organizations of Colombia, Ecuador, Peru, Bolivia and Brazil. With that organization we are trying from the broadest possible base to strengthen and support our member organizations and establish a close and lasting relationship between them. We are also trying to fight for the rights of the indigenous people of the region through united action.

The Coordinadora is trying to achieve the active participation of official organizations such as OES, the UN, OIT, UNESCO and others, to keep them informed and to present to them our legitimate claims, to inform national and international opinion about the grievances of the indigenous populations.

I would like to extend greetings to all those organizations whose struggles have been recognized by the award of this prize. I would also like to extend greetings to all those organizations who by their solidarity and help have contributed selflessly to the struggle for the re-establishment of the rights of our people. I want to tell you that their effort in the European countries has increased the understanding of the problems our people face and we hope that in this way we can all take part in the search for peace and social justice for humanity.

I also want to greet and at the same time thank the ecological movements in the world who have recognized the vital importance of an ecological solution which we have always advocated especially in view of large multinational companies destroying resources in different parts of the world. Because of this, we ask them to continue their fight and continue to help us with their strength and enthusiasm.

The Sarawak Natives' Defence of the Forests

MOHAMED IDRIS AND KHOR KOK PENG
SAHABAT ALAM MALAYSIA, SARAWAK
(FRIENDS OF THE EARTH, MALAYSIA)

December 1988

THE BLOCKADES SET UP by thousands of native people in Sarawak in 1987 captured the world's imagination. They stopped logging activities in most parts of the two major timber districts of Sarawak, Baram and Limbang for several months. Although the blockades were dismantled in October 1987, new blockades again appeared between May and November 1988. At present there are reported to be between three and five places where blockades have been set up.

What drove these peace-loving people, especially the forest nomadic people, the Penan, to such action? It was the logging of their forests, on an unprecedented scale, by timber companies. Most of the logs end up as furniture, house materials and packing materials in the industrialized countries, especially in Japan.

The indigenous people had been manning the blockades in continuous 24-hour cycles, braving the sweltering tropical sun in the daytime and the cold mountain winds at night.

Many of the Penan had walked many days from their homes in the deep forests to keep vigil at the blockades, leaving behind their families to fend for themselves. They have set up temporary wooden shelters near the barricades, taking turns to rest or to cook.

The Penan are the original inhabitants of these 150 million-year-old Borneo forests, the oldest in the world. Being hunters and gatherers, they do not have permanent settlements, unlike the other tribal communities. Under state laws they have minimal land rights, and they have been fighting a losing battle for the past five to seven years against the timber companies which won concessions to log the forests. The plant, food and wildlife resources of the Penan have progressively disappeared and their water systems are clogged with silt, debris and chemicals from logging activities.

'Before anyone else, and long before the timber company came, we were already on this land,' said Along Saga, an elderly Penan in his mid-fifties. 'This is the land of our forefathers, and their forefathers before them. But now with just a few years of logging these centuries-old forests are almost finished.'

SAM sent a memorandum to the authorities on behalf of the indigenous people, calling for the cancellation of all timber concessions which have not started operations, for fair compensation to those whose lands and trees were damaged by logging, and for the recognition of their customary land rights.

In particular, the people want a review of the land laws to increase the area of their protected communal forests, and to safeguard their rights to customary lands and forests from the encroachment of loggers and other outside forces. These demands were now, for the first time, backed up by the blockades and won international recognition. But the indigenous people are up against formidable odds.

At the heart of the conflict is the control and use of the world's oldest and richest rainforests. The Borneo jungle contains the greatest diversity of plant and animal species existing anywhere on earth, holding as yet undiscovered scientific secrets and playing a vital role in balancing the world's natural ecology. In the past twenty-three years between 1963 and 1985, a total of 2.8 million hectares (or 28.217 square kilometres) of primary forests were logged. This is equivalent to thirty per cent of the total estimated forest area in Sarawak which is 95,232 square kilometres.

As of December 31 1984 (the latest statistic available), some 5.8 hectares of forested area were under concession. This is equivalent to sixty per cent of Sarawak's total forest area. In other words, three-fifths of Sarawak's forest was then licensed out for logging. In the run up to the State elections in April 1987, it was publicly made known that most of these concessions are owned by prominent politicians, their families, friends and political allies.

At a press conference on April 9 1987, the Chief Minister, Datuk Patinggi Abdul Taib Mahmud, revealed that he had frozen twenty-five timber concessions totalling 1.25 million hectares belonging to relatives and friends of the former Chief Minister, Tun Abdul Rahman Yakub.

In response, Tun Abdul Rahman Yakub also revealed to the press the names of politicians, their relatives and associates connected to Datuk Patinggi Abdul Taib Mahmud who controlled some 1.6 million hectares of timber concessions (*New Straits Times*, April 10 1987).

It can be seen that between them they control some 2.9 million hectares of forest concessions, which is more than half of the total area of forests licensed out to be logged. If one were to include the rest of the existing timber concessions, one can say that the majority of the forests of Sarawak must be given to politicians and their supporters.

Compare this to the amount of Communal Forests Reserves which have been granted to the natives in the past. Since 1968, the area under Communal Forest had shrunk from 303 square kilometres to only 56 square kilometres in 1984. In the space of a few years, the Communal Forests have been drastically reduced by eighty-two per cent, thus depriving the native communities from having access to forest produce in the vicinity of their longhouses. Natives obtain their basic necessities (such as wood for their boats, houses, farm implements, and medicines when they are sick) from the forest, which is their only source of survival.

To our knowledge, eighteen communities in the Limbang and Baram areas have applied for Communal Forest Reserves and none of them have been approved. This clearly reflects the negative stance towards the native communities in their applications for Communal Forests (a right upheld in the Sarawak Forest Ordinance). Yet, the same forests are often given to timber licensees.

According to senior foresters, large-scale hill logging operations in Sarawak are already causing floods, siltation of rivers, turbidity of upstream river water, and a reduction in the aquatic and wildlife population. An FAO study has shown that in logging, as much as fifty per cent of the residual stands in the forests may be damaged, and the surface soil may be destroyed, when up to thirty per cent of the ground surface is exposed. Residual stands are the remaining trees which are not wanted and hence left standing during logging operations.

The study also says that it will take more than forty years for such disturbed forest to recover. It is disturbing to note that in Sarawak after twenty-five years, the same area of hill forest is logged again in rotation under the Selective Felling System.

According to FAO forestry officials, the three important causes of logging damage in Sarawak are felling, extraction and road building. A World Wildlife Fund (WWF) report states: 'The extraction process leaves an average of thirty-four per cent of a forest stand open, even at the prevailing rate of seven trees taken out per hectare. Thus, hill forest logging in Sarawak removes about forty-six per cent of the natural cover.'

Poisoning also leads to the long-term contamination of the environment. The residues of these poisons have long-term effects on aquatic and wildlife, the soil, vegetation and health of humans living in these areas.

According to Professor Soepadmo of the University of Malaya, 'Along with this loss of tree genetic resources will go hundreds of other plants which originally characterized the complexity and species diversity of the primary forest. The disappearance of these plants will also mean the loss of habitats for various animals inhabiting the forest.'

A joint report of World Wildlife Fund and the Sarawak Forest Department says that while 'hunting and gathering is becoming more and more difficult for all rural Sarawakians, it is the nomadic and semi-settled Penans who are most affected, as selective logging and forest clearance reduce their supplies of both forest food (sago, wild animals, fruits, vegetables, etc.), and of rattan and other non-timber forest products that have both domestic and cash significance. In addition, their water supplies and supplies of fish, etc., may be adversely affected by soil erosion and siltation over which they have no control.'

It can be seen from the above detailed scientific studies cited, that deforestation in Sarawak has led to serious consequences on the ecology, food resources and livelihood of the native peoples of Sarawak.

Ever since the timber companies operated in their areas, the communities have made repeated appeals to the timber companies and State authorities to help them in their distress. They have complained that logging activities have damaged their farmlands, their water resources, sacred ancestral burial grounds, fruit trees and other forest resources. Wildlife has disappeared and fish has become scarce, while their children have fallen ill consuming water from rivers and streams which have become contaminated. They have written numerous letters appealing to the Chief Minister, the Forest Department, the State Secretary, the President, the District Officer and have even lodged police reports regarding damages to their property. In many cases there has been inadequate response. In the case of the Penan communities, they have repeatedly petitioned the Sarawak Government and the timber companies to stop the destruction of their lands and their livelihoods. They also met local authorities several times but to date, despite the promises to look into their problems, nothing has been done.

The native peoples are very loyal, peace-loving and law-abiding citizens. By tradition, they have always considered the Government as a benevolent father who looks after the needs of his children. They are not 'wild', 'primitive' or 'uncivilized'. Neither have they been misguided, tricked or 'led' by foreigners. They just want to live their way of life in peace without any disturbance. What have they done to deserve such a fate? What crime have they committed when they are just protecting their forests, lands and survival?

Let us now examine the scientific facts on who is responsible for forest destruction, the shifting cultivators or the timber industry.

In 1985, 270,000 hectares (or 2,700 square kilometres) of primary forests were logged in Sarawak. This was equivalent to 2.8 per cent of Sarawak's forest area. Should logging continue at this rate, another twenty-eight per cent of Sarawak's forests will be logged in the next ten years and what remains behind will be virtually wasteland.

Shifting cultivation is based on a system of land rotation with a period of long fallow. According to Dr S. C. Chin, an ethnobotanist at the University of Malaya who has spent the last twelve years conducting intensive research on shifting cultivation among the native communities in Sarawak, a family of shifting cultivators clears and plants an average of about five acres annually. This means a total of 180,000 acres for the whole of Sarawak (given the current estimate of 36,000 shifting cultivator households in Sarawak). The rest he says will be from agricultural land that has been fallowed.

'Therefore to say that shifting cultivation destroys 100,000–150,000 acres of forest annually (as is frequently alleged) . . . is completely misleading and can only be intended to confuse,' says Dr Chin.

Thus the amount of primary forests opened for shifting agriculture each year (3,600 hectares or 9,000 acres) is only a small fraction of the 270,000 hectares logged by the timber industry in 1985. Moreover, scientists worldwide have increasingly acknowledged the fact that shifting agriculture is a highly complex system, 'stimulating the structure, functional dynamics and equilibrium of the natural ecosystem more than any other agricultural system man has devised'. According to Dr Chin, 'shifting cultivators know better than anybody else that their system of cultivation depends on the successful regeneration of the forests.'

It can be seen that the rapid destruction of the State's forests has been wrongly blamed on shifting cultivators, whilst the real factor responsible, the timber industry, has hardly been focused on for its role in depleting forest resources. It is clear that the timber industry has far more capacity to destroy Sarawak's forest, and has used this capacity with increasing effect in recent years.

What is most fundamental to the natives is the question of land rights. The natives of Sarawak have lived on their lands for hundreds of years, even before the existence of the nation state. In the case of the Penan who are the true forest dwellers, their existence in these forests would have stretched back to the dawn of humankind.

Their reverence for the land meant that it could not be bought or sold. This principle was enshrined in *adat* law which has legal, moral and religious aspects. Under *adat*, the community as a group exercises rights to the land. This concept of justice and morality *vis à vis* land is a living concept among the natives of Sarawak.

Under the Sarawak Land Code this customary right to land practised by the natives is recognized and enshrined as a basic principle. Thus, legally and morally, natives have a right to their lands (on which they have lived for generations), and they want this right to be fully recognized and guaranteed by the State. One should not force them to leave their forest lands and resettle them elsewhere, all in the name of development. In the case of the Penan, it is tantamount to cutting off their life support system.

What the natives want (as expressed in a resolution of July 1987 of sixty-one communities in Baram and Limbang) is a kind of development suitable to their needs. 'Development to us means: (a) the recognition of our land rights in practice; (b) the cessation of logging in our lands and forests so that we can continue to live; (c) the introduction of a clean water supply, proper health facilities, better schools for our children. This is the kind of development we want. Why don't you give us this development and progress?'

The Treatment of Torture Victims: The Work of the Rehabilitation and Research Centre (RCT) in Copenhagen

INGE KEMP GENEFKE

December 1988

IN THIS PAPER I WILL DESCRIBE (1) the aim of torture, torture methods, and after-effects. Methods of physical torture will be dealt with to begin with, and subsequently those of a psychological nature. It is a hard and grim paper; it cannot be presented lightly or easily. It is certainly not my aim to do so. In our opinion it is impossible for anyone to perceive what torture is really all about unless they have experienced it. The only people who understand the real meaning of the word are the victims who have actually been tortured. We, who have not been victims, of course have acquired an amount of knowledge about methods of torture and the results of these methods. But the fear and anxiety which results from being present in a torture chamber, where trained tormentors, six, eight or ten in number, try to destroy the victims both mentally and bodily, the fear and anxiety which they experience, we simply do not understand.

After this I would like to explain (2) how we, at the International Rehabilitation and Research Centre for Torture Victims (RCT) have provided help for torture victims who live in exile. I shall conclude with (3) how we at the RCT imagine that this help is to be expanded and developed in the future, both nationally and worldwide.

1) The aim of torture is not, as is generally understood, to receive information from the victim. No, the general aim of torture is to destroy a human being, destroy its personality, identity; it could be said to destroy the soul of a human being. This is a great and important revelation which is of enormous help in our work.

And now for the methods of torture. First the physical methods.

When discussing torture this may be done in very academic terms. One definition of torture is from the so-called Tokyo declaration from the World Medical Association (WMA) in 1975. Here the definition of torture is as follows: 'the deliberate, systematic or wanton infliction of physical or mental suffering by one or more persons acting alone or on the order of any authority, to force another person to yield information, to make a confession or for any other reason.' . . .

In order to explain to you what torture really is, I prefer to describe the following scenes. [In her presentation Dr Genefke used photographic slides at this point.]

Big, strong police or military people enter the home, destroy the home, smash it up, hit the wife, perhaps even rape her, hit the small children. It is not difficult to imagine that children having had such devastating experiences may well suffer for the rest of their lives, if they are not helped. At the centre we have children who have experienced their pets, dogs or cats, being killed in front of their eyes. The father may have been beaten unconscious.

Subsequently, the torture victim is brought to the torture chamber. The first days they go through raw, unsystematic torment, being hit and thrashed unmercifully all over the body. It is unnecessary to mention that they are not allowed to go to the toilet, that nothing is given either to eat or drink. The next picture shows that the women have their clothes taken off altogether. They are humiliated, scorned. You can see that some of the women have blood from menstruation trickling down their legs, they are called pigs and swine and other humiliating names. This is followed by more systematic physical torture. This shows what we name 'the grill'. It consists of an iron framework which is heated electrically. It may be well understood that the person lying on it will sustain very deep and severe burn injuries, which may also be seen in the next picture. This burn injury situated near the rectum is naturally a very painful wound, but its location is also extremely humiliating.

Already now a picture of the mixture of physical and psychological torture taking place is revealed. Torture is always a mixture of physical and psychological methods in which humiliation plays an essential part.

There is another case of a journalist examined by myself, suffering from burn injuries. He had been injured with third degree burns caused by an iron bar on thirty-five different parts of the body. He told me that he sensed the smell of burning flesh, then he fainted, and was awakened by cold water being thrown in his face. This was also a case of a severe combination of physical and psychological torture. They started to burn his left hand and they informed him that they knew he was a journalist and would soon continue burning his right hand, so that he would never again be able to work as a journalist.

Another scene shows a watercolour made by the same torture victim. It depicts submarine torture. A doctor is present, his duty being to observe that the pulse does not stop or that the nails do not turn too blue. The idea of the torture does not aim at the death of the victim. The victim is supposed to go on living, thus—if the pulse becomes irregular or the matrix of the nails turn too blue—this mode of torture is discontinued for a short while.

A further case reveals electric torture inflicted on a female, the form of torture we call picana. The electrode causes severe cramps, causing unbelievable physical pain. At the same time the humiliating situation the female is subjected to will be realized, naked as she is in the company of these men. This is very severe psychological torture.

Finally as an example there is the case of a person from the Philippines. A group from the RCT went to Manila, where we at the University of Manila instructed doctors, psychologists, health workers and lawyers, all coming from the Philippines. We were handed this picture by a doctor who had seen this person. He had a nail hammered into the left side of the brain; the person in question was unable to make any accusations being lame on the right side and unable to speak.

These were the physical forms of torture, and we regard them as very, very severe, but I would like to make it clear that the worst form of torture is psychological. The physical after-effects of torture are in accordance with the method by which torture has been inflicted. After having been subjected to falanga torture, one may have difficulty in walking, after being hung up one way or another, one may achieve joint difficulties, etc.

The mental after-effects of torture are always the same; you feel anxious and depressed, you have nightmares about the horrors you have experienced, you cannot sleep at night. You have difficulty remembering, you have difficulty in concentrating. You are always tired and suffer from headaches, and you may have sexual problems. Worst of all you feel your personality has changed; whereas you used to be strong, happy and optimistic, you are now depressed. Previously you were extroverted and active, now you are introverted and isolated. Previously you had faith in yourself as a strong person, now you feel weak. You have also lost confidence in other people. Previously you were well-balanced, now you easily get irritated and angry. I want to stress that at the RCT we consider all of these problems to be completely normal reactions of normal people to something which is incredibly cruel and perverted and which normal people cannot understand, i.e. torture.

Isolation in very small cells, perhaps measuring two by one-and-a-half metres, which either are kept quite dark or fully lit, is a major component of the psychological dimension of torture. Perhaps also one has a hood over the head, and food is given at varying times of the day, so one is deprived of any sense of time. Obviously one is not allowed to visit the toilet, and no possibility of washing oneself exists. Under such circumstances hallucinations may arise, and the torture victim has a sensation of going out of his mind. Obviously the aim of torture is to create this fear of losing one's normal senses.

It is a major component of torture, that human beings are humiliated to such a degree that they become isolated. They are told that the things they have experienced here they will never be able to tell any other human being. Thus, an extreme feeling of loneliness and isolation is created for the torture victim.

Threats and witnessing torture is the very worst of all, naturally if the person in question is dear to you, friends, partners in marriage, one's children. At the centre we have children that have been tortured, witnessed by the parents. It is insufferable to be defenceless when listening to the cries and prayers for the torture to stop. This is what gives rise to the worst nightmares and this is what leads to the most serious after-effects. I do not believe that anyone among us is really capable of understanding or realising the fear and sorrow the victims of torture go through when exposed to this form of torture.

Sexual torture takes place in a variety of ways, in which humiliation always enters as a prominent component. All victims are exposed to sexual torture....

Let me repeat: the aim of torture is to destroy a person as a human being. This is just what we have to understand when we experience this sort of horror. It is the identity and soul of the person one has sought to destroy. This is just the most cruel part. In reality it is more evil than murder.

2) What assistance can be given to people who have been exposed to torture?

First, I would like to describe the facilities and location of the RCT and then the principles according to which we help the torture victims who come to us.

The RCT is housed in two buildings close to the University Hospital in Copenhagen and covers an area of approximately six hundred square metres. There are thirty-two staff members at the Centre including health staff such as doctors, psychotherapists, nurses, physiotherapists, social workers, and administrative staff. Interpreters, various specialists, and consultants such as dentists regularly come to the Centre. We first of all give out-patient help to victims who have obtained asylum in Denmark. Hospital help is attached to the centre as the University Hospital has placed two beds in the neuromedical department at the disposal of the people for whom hospital admission is deemed necessary. This may be for torture victims who need to be operated on, or who need to have extensive examinations performed; or, for example, for those suffering from severe depression.

In our help to victims we have reached some fundamental guidelines which we always follow. We have laid down these guidelines in our work with the torture victims, but also with doctors and other health staff from Third World countries who simultaneously with us try to help torture victims in their own countries.

The first principle is that it is important when performing examinations and treatment to avoid methods which may bring back memories of the torture. You may think this is easy to avoid but in reality there are many things in our daily life and in our hospitals which may remind the torture victims of prison experiences and torture. An ordinary blood test which does not mean anything to the rest of us may bring back horrible, evil memories of people who have been soaked in their own blood or who have seen friends or other prisoners with blood running down them. An ordinary electrocardiographic examination of the heart where electrodes are strapped on your arms and legs is nothing for the rest of us, but if you have been tortured with electricity it may be an examination which cannot be carried through at all.

There are also other things one has to be aware of: small rooms may bring about anxiety if the victim has been isolated in small cells for months. The same applies to lifts, or to closed doors, and bright lights. It goes without saying that a gynaecological examination of a woman who has been raped several times may be a very straining experience. Nearly all victims have been naked during the torture so if you ask a person to take off his/her clothes it can also be a very terrible experience. Often examinations may have to be postponed or performed very slowly and calmly, with full explanations given about why this and that must be done. . . .

This may sound complicated and difficult. In practice it is not. The main consideration is knowing what the torture victim has been through and acting accordingly, with appropriate sensitivity and respect. Furthermore, the cultural aspects are also taken into consideration.

The second principle in the treatment of torture victims is that both mental and physical effects are treated simultaneously. The mental effects are treated by long talks about the torture which the victim has been exposed to. You try to make the torture victim talk in detail about the awful things that have happened and the feelings which the torture victim felt during and after the torture. It concerns handling the sorrow and anger situated deep in the mind. Naturally these are very difficult conversations in the early stages. They take place twice a week as out-patient treatment. It is often very difficult for the victims to talk about these deeply humiliating experiences. However, it is a help that you can use the dreams of the victims. You can ask the victim what he/she dreamt about the previous night, and since it nearly always concerns the prison and torture, it can be used as a basis for conversation, discussion, and counselling. It is often possible to help the victim get better by being able to follow a development in the dreams.

Simultaneously with these conversations we examine the victim thoroughly, medically and physiotherapeutically, i.e. especially with regard to joints and muscles. Almost all torture victims have distinct physical pains and therefore we also treat everybody who comes to the centre with physiotherapy in a small clinic. We are careful in this treatment not to inflict any kind of pain and overall we try to avoid anything in the treatment which may recall very bad memories.

The third principle is that you do not only help the torture victim but if there are co-habitants or spouses and children we also help them and consider the social conditions; i.e. we try to help during the period in which the victims come to us. As stated earlier, it is not only the torture victim that the torturers have struck but often the entire family. When a family member feels bad it naturally rubs off on the rest of the family. This person has often had to live without his/her father, mother or both parents for long periods. Perhaps the children have been left to other family members, have missed their parents, have not understood why their parents have suddenly left them. Maybe the children have also been taken to prison, maybe the parents have been displayed to the children in a severely tortured condition. Therefore it is important that we offer help to the whole family, naturally on a voluntary basis.

3) How do we expand and develop this help further both nationally and worldwide? I am briefly going to explain how we at the RCT imagine this can be done. Many of the areas I describe here are already worked on at the centre.

It is important to understand that helping torture victims is a relatively new field. Therefore it is necessary to collect more experience and it is necessary to carry out research just as it is necessary to teach. The conditions for being able to do this are that there is a relatively large centre which can function as a nucleus for research and teaching. The best thing would be if there was such a centre in all countries.

TEACHING

It is important that the knowledge and experience which we already have today is passed on to health staff under training; health staff who are particularly in contact with refugees, including torture victims; and health staff in countries where torture takes place. This is the reason why the RCT has worked internationally to realize these plans. At the World Health Organization-sponsored meeting in 1986 in the Netherlands where the RCT was co-organizer, a proposal put forward by the RCT was adopted. The proposal recommended teaching of health staff on how to help 'victims of organized violence'. This has been sent as a recommendation to the European governments. The work is followed up in co-operation with the WHO and the UN.

At the RCT we have worked out educational material, films and textbooks for health staff. Educational material has already been completed for medical students, nurses and physiotherapists and is now being prepared for all professional groups within the health sector. The plan is to collect this material in a textbook.

In Denmark, the RCT has contact with health staff in the primary health service who are particularly interested in helping torture victims. A network of general practitioners, psychiatrists and psychologists taught and supervised by the RCT has been created.

RESEARCH

The object of research at the RCT has been:

1. To prove the purpose of torture;
2. To describe torture methods, physical and mental;
3. To describe the consequences and effects of torture, both physical and mental. It is important to objectify the effects of torture with a view to being able to prove effects of torture and as a help to the important final objective;

4. To find the best help for torture victims.

Items one and two have today been achieved.

At the RCT we have many current research projects. Here only a few examples will be given:

▷ Evaluation of the condition of the shoulder capsules after suspension from the arms. The pathological changes in the shoulder capsules depend on the way the torture victim has been suspended. A particularly evil way is with the hands tied behind the back. Through these examinations you partly find out what injuries have happened, possibly pathognomonic to prove torture. Furthermore you achieve a better understanding of the extent of the injury and thereby which rheumatological treatment will be the most suitable.
▷ An important and painful effect of torture is sleep disturbances. A research project is aimed at continuous monitoring during sleep. Hereby the disturbances in the sleep pattern which are developed in people who have been exposed to torture, are studied. . . .
▷ A special research study is aimed at unveiling the involvement of doctors in torture. At the RCT we believe that torture cannot be carried out in the world to the present extent without the participation of some members of the medical profession. A prospective study is aimed at disclosing how many of the torture victims who are treated at the RCT have been in contact with health staff in connection with the torture. Already at the present time the figures show a frighteningly high percentage.

THE RCT DOCUMENTATION CENTRE

The RCT Documentation Centre was established and officially inaugurated in 1987. Its objective is two-fold: first to collect and register literature on torture, the effects of torture, and on the treatment and rehabilitation of torture victims; and secondly to disseminate this information to health personnel and researchers in various institutions and organizations all over the world involved in work with torture victims.

This is the first attempt to compile all the available information under one roof. The RCT Documentation Centre is the first institution to strive for implementation of the World Health Organization proposal from April 1986, calling for the centralization of literature on organized violence.

In 1987, the necessary computer equipment required for a data base was purchased and a highly sophisticated data base program for bibliographic data was installed on a personal computer. Since the official opening the Centre has offered searches from its own data base, print-outs of references and, to some extent, lending and photocopying services. Its services are at present primarily intended for researchers and health personnel involved in work with torture victims. A Newsletter, available to colleagues working with the same problems, is now being published, at present four times a year.

INTERNATIONAL ACTIVITIES
From spring 1988, the objectives of the Centre were changed so that international activities have been separated, with their own function and budget, with the following objectives:

1. To provide information about torture, and on the rehabilitation of persons who have been subjected to torture.
2. To support rehabilitation activities.
3. To contribute to and support research on the nature and extent of the consequences of torture, with regard to prevention, treatment and rehabilitation.
4. To contribute to and support the teaching of health staff in the examination and treatment of persons who have been subjected to torture.

The RCT has created together with the French Centre for Torture Victims (AVRE) and the American Centre for Torture Victims in Minnesota, an International Foundation and the international work at RCT is related to these international activities. . . .

In preventing torture and in helping its victims, I believe the role of health personnel to be of crucial importance. It is therefore imperative to have knowledge about torture communicated to health personnel within the world medical organizations in order for them to participate in the work of rehabilitation.

Tribal Peoples and Survival International

STEPHEN CORRY

December 1989

SURVIVAL INTERNATIONAL IS A MOVEMENT, twenty years old this year, which helps tribal peoples to protect their lands and ways of life from destructive outside interference.

We all now know what the outside world is trying to do to tribal peoples. Basically, outsiders want their land. Having exhausted its own resources, the so-called 'developed' world is now running all over the globe tearing the minerals, oil, wood and so on from areas which were once remote and thought to be useless. Or, which is often worse, governments are pushing poor people to colonize these places to avoid fair land distribution elsewhere. They are carting the poor off to land which the rich do not want.

More often than not, these are the territories of the world's remaining 200 million tribal people.

Tribal peoples depend on these lands like no one else does. They get all their food, medicines, building materials, and spiritual meaning from what is around them. Invade the fragile Amazonian forest and you scare away the game over a huge area. The Indians cannot hunt, so at best are forced to work for outsiders as labourers or prostitutes. They get sick from diseases they never had before. They are quickly destroyed. Or, put a mine on an Australian sacred site and you put a knife straight into the heart of an Aboriginal people's heritage. You cut off roots which nourish their souls and which may be 40,000 or 50,000 years old.

Tribal peoples, indigenous peoples—call them what you will (and there are important problems with all the possible terms)—are the victims of our greed for resources, coupled with a blatant and extraordinarily deep racism which says that they are backward and primitive and must catch up with the twentieth century or perish. An often heard refrain is that we cannot hold up progress because a few naked tribals have lived on top of these oil fields since before recorded history.

How short-sighted, indeed how primitive, this view is. Apart from our technological strength, which in any case may well be short-lived, there is clearly nothing superior about our own way of life. Tribal peoples have their own technology, their own medical systems, their own education and so on, and these work. They are not stupid. If they did not work well, they would soon change them.

These facts are self-evident to all who have brought an open mind to the privilege of being a guest of self-sufficient people.

One of the most common criticisms levelled at Survival over the years is that we want to keep people as they are. A 'human zoo' where cultures are preserved. There is no truth whatsoever in this, and the allegation itself betrays several fundamental misconceptions about tribal peoples and cultures.

The notion that somehow or other 'our' culture is the most advanced and that everyone else wants to become like us is nonsense. But so is the idea of being able to 'preserve' cultures. All cultures, all peoples, alive in the world today belong to the present, none are anachronistic. None are remnants of the past, destined to perish through the passage of time.

Why do people think that tribal peoples' cultures are inferior? In the Third World there is a truly startling paradox when you compare these people with most others. Take Amazonia, for example, where traditional Indians live well in comfortable dwellings—warm at night and cool in the day. They eat well—a varied and healthy diet. They live in a close community where loneliness is unknown. And they do it all on three or four hours of work a day, or less, and have plenty of time for playing with their children, for contemplating philosophy, cosmology and religion, and for externalizing whatever answers they find through profound rituals which make many of our own seem shallow and meaningless.

Compare this life with the lot of the Third World poor who supposedly benefit from 'civilization' and who are by and large growing poorer daily in spite of the billions in aid. Their children are working fifteen- to sixteen-hour days. They are badly nourished. Serious disease is rife and Western technological medicine largely unobtainable. Infant mortality high, life expectancy low, alcohol and drug abuse common—social breakdown is often the norm in the shanty towns where life comes and goes on the cheap.

The resource extraction, the dam building, most of the so-called 'development' going on in these countries benefits neither the tribal peoples whose lands are destroyed in the process, nor the vast majority of the nation's citizens, the poor and needy. The ones who profit are, of course, the wealthy—and the governments are always wealthy—and the foreign companies.

And here it is important to point out that it does not seem to matter much which political side—left or right—any particular government claims allegiance to. At no place on the political spectrum do the rights of tribal peoples seem to count for much. Neither is the problem confined to Third World dictatorships or poor countries; it is not just the result of national debts imposed by rich countries.

The situation for tribal peoples in the Philippines is now actually worse than it was under the dictator Marcos and their communities are being regularly attacked by the armed forces. Or take the case of Australia where Aboriginals, the lucky ones that is, were kicked out of their desert homeland to make way for British nuclear tests. The unlucky ones were left there to die. Or consider Canada, where even as I speak, the countries which make up NATO are deciding if they will set up their low flying war games which are destroying one of the last traditional Indian societies left in North America.

Survival International began work twenty years ago by putting a lot of effort into researching concrete projects directly with tribal peoples. We had seen that there was plenty of money going into projects in the Third World, but that little or none was getting to the people we were concerned about. Funding agencies did not have enough time or resources to reach remote areas, and they lacked background and, often, interest in the subject. During the 1970s, a large part of our work was put into finding good projects and then matching them up with sympathetic non-governmental funding agencies. This worked well and it was not too long before many of the funders were developing their own programmes with tribals and our intermediary role could be phased down.

This is still the case, and nowadays we have many more campaigns and educational programmes than field projects. But we do still back some, carefully selected, individual projects. Where possible these are designed and administered by the people themselves.

Those which we do support are generally concerned with health, education and marketing. These are the three key areas where outsiders complete the dispossession of tribes which started with land invasions. And it is in these sectors that many who are actually well meaning towards tribal peoples show their unwitting cultural arrogance and paternalism (their 'ethnocentrism' to use the jargon).

The way indigenous people treat their sick, teach their children and gain their livelihood are constantly belittled by outsiders. Indeed, many believe that they do not actually have any medicine or education.

We have supported several projects to put these central elements of all societies back into their hands once again; for example, teaching tribal paramedics the basic applications of Western medicine, demystifying it and enabling it to be used in conjunction with native, often shamanic, healing. Money has been used for training courses for tribals and to supply them with basic drugs. More straightforwardly, we have also supported vaccination programmes for groups, such as the Yanomami of Brazil as well as those in Venezuela, who have only recent contact with outsiders and now risk devastation from the new diseases being brought in.

Similarly with education—we have supported the training of indigenous teachers and the publication of school books in the tribal language, so that outside teachers, who are invariably racist, who use only the national language and who often punish native children for speaking their own tongue, could be replaced by teachers from the indigenous group itself. One of these projects, with Amazonian Indians in Peru, brought a dramatic increase in literacy and school attendance in only a few months. Another, in Australia, eventually shamed the government into directly funding the Aboriginal-run school itself. The purpose of these schools, which complement traditional, non-school, education, is to teach children about how to cope with the unfamiliar world without getting cheated. They will not learn this from unsympathetic and uninterested teachers from outside.

The projects we have supported in the economic sphere have often been more complicated, and many have not proved immediately successful. One which apparently failed, actually brought surprising and very positive developments in its wake. This was one of the first projects we supported and was a response to a request from the remnant survivors of the Amazonian rubber massacres earlier this century. Tens of thousands of Indians had been enslaved and killed with extraordinary brutality. In one group in Colombia, the Andoke, only about one hundred people survived out of a population of 10,000. They were still enslaved in permanent debt to the rubber baron when we met them in 1974.

They asked for money to get them out of his clutches and to help them set up a rubber co-operative under their own control. We gave them the capital. Their scheme only lasted a short while and then collapsed, but through it the people found their pride and independence again. They rebuilt their traditional longhouses, and they reaffirmed their own culture which depended on the profound symbolism contained in those longhouses and the religious meetings and dances which could not be properly performed without them. The Andoke's days seemed numbered when we visited them fifteen years ago. But against all expectations, they are now a thriving independent Amazonian community.

Survival has shifted its emphasis over the years. Nowadays, it is primarily a campaigning and educating organization. It is a financially independent, non-governmental and non-profit making organization funded by our members and by the public, with smaller amounts coming from private trusts and foundations. No single donor gives more than three or four per cent of our total income and we do not have money from governments. Our five national offices, in the UK, US, France, Italy and Spain, rely heavily on volunteer staff backing up small teams of professionals. We have grown quite quickly over the last few years and are looking to expand our work all the time. This depends, of course, on more money being available. We currently have about 8,000 paying members in over sixty countries, though our materials are actually sent to about 30,000 organizations and people, many in the Third World. We regularly publish in six European languages and Japanese, though most of our material is in English, French, Spanish or Italian.

In a year, we usually handle about fifty specific cases covering about twenty-five countries; we emphasize South America and Southeast Asia but also work on Australasia, the Indian subcontinent, parts of Africa, the Pacific and North America. In the past we have given some support to the Sami people of northern Europe but, in general, we give priority to groups who have only recently come into contact with outsiders and who have no organizations to represent them. Generally, it is they who have the most to lose.

Survival focuses its attention on to deliberately narrow aspects of the injustice which tribal peoples face. We stress continually that they have rights to the lands they live on and use. This is their land. They may not have concepts of land ownership, they may not have paper titles, they may not speak the national language or even have any idea they form part of a nation state. But none of this alters one jot the fact that the land is theirs—by moral right and by law. For over thirty years, the international law on tribal populations has stated unequivocally that they have full legal right to their land. Governments which flout this, as they practically all do of course, are acting illegally.

Survival puts a lot of weight behind this question of land. We believe that this is where the central battle lies. If proper land rights can be secured—and they should be full ownership rights held in common by the people, not just reserves—then tribes will be able to choose themselves how they are going to adapt to the changing world.

We believe that governments must acknowledge that tribal peoples have this right. We believe that governments are not actually going to recognize this unless they are made to and that they are not going to be made to other than by force of public opinion. We are looking for long term solutions—ideally very long term solutions—so that tribal land is free from invasion for once and for all time.

This may sound unrealistic and utopian and of course it is unrealistic and utopian. There is no point in working for an ideal unless you aim directly for it. We are not traditionalist. We are pushing for a profound change in attitudes. People must realize that the effects of their racism should not be allowed to kill people. We may all be racist on a deep level, including tribal peoples themselves, but it is the effects of that which bring tyranny and violence, not perhaps the racism itself.

We are pushing for a recognition of human values over economic and political expediency; for an acknowledgement that many of our technological answers to life's problems are in fact failing and that we must listen now, before it is too late, to some very simple and actually very obvious concepts which are encompassed in many tribal peoples' attitudes to the world. Here we must be careful not to fall into a 'noble savage' trap. They have not got all the answers and they suffer in their traditional lives, the same as everyone does and from much the same things. They do not spend all their time communing with the natural world like a group of exotic nature freaks or naked, romantic poets. They are prey to the same vanities which afflict all people.

But they do live by an evident fact; people do not exist outside of the environment. People and the environment are one thing. Tribal peoples know this—we have forgotten it. And it is with considerable dismay that we see that even some of the environmentalists have forgotten it. They project a short-sighted view of ecology; an image of 'destructive man' versus 'stable nature'. This is actually a very narrow idea which does not really hold up when scrutinized. But it has been applied with disastrous consequences for tribal peoples. Whole tribes have been annihilated because their land has been turned into National Parks. People who have lived there for thousands of years are now denied land rights and forbidden to hunt because we want to preserve animal species for our own study and pleasure. Yet, of course, it is not they who have brought the animals to the edge of extinction.

Survival is currently pressing for proper land rights to be recognized in spite of National Parks in Ecuador, Malaysia, Panama, Botswana and Peru. In Botswana they are planning to dispossess Bushmen because the government claims they are disrupting wildlife. Can you imagine, having destroyed practically all the animals, they are now pretending that the Bushmen, who have lived there longer than anyone, are a danger to the game?

Survival has successfully argued at the International Whaling Commission for many years that the Inuit (better known as Eskimo) of Alaska be allowed to hunt whales for food. Survival representatives there have always been Inuit themselves.

In seeking long term solutions to all these problems we have seen that only worldwide public opinion is going to bring any profound impact. When many shareholders are genuinely outraged that their companies are destroying Australian Aboriginals, when development agencies know that many taxpayers will not tolerate their funding genocidal projects in India or Indonesia—then the situation will change. Public opinion nowadays would not allow a reopening of the slave trade from West Africa to the New World. Public opinion eventually threw out Marcos from the Philippines and look what it is doing in Czechoslovakia and East Germany today. Public opinion, we believe, will stand up for tribal peoples' rights and will bring profound changes to how they are treated. We are optimists actually.

Survival is doing two things; catalysing and crystallizing. First, it is trying to catalyse this public opinion through its educational work; publishing newsletters, reports, books and exhibitions, giving talks and slide-shows, organizing school projects, and so on.

We publish a series of popular periodicals and occasional more technical reports. As well as telling people how tribes are being oppressed and killed, we want to publicize the beauty of their ways of life and explain how much they have already given us and how much more they could give; how so many of our foods and medicines come from them, from potatoes to muscle relaxants vital for major surgery, and how within their knowledge of the tropical forests, may lie cures to some of our most lethal illnesses.

Second, and most importantly, Survival is trying to crystallize this opinion which we and many other groups are raising, and which is growing anyway, to focus it into organized and sharp campaigns, well targeted, using proven methods and seeking realistic, more immediate, goals. We get media attention, use our consultative status at the United Nations and similar organizations, make constant representations to governments and multilateral development banks, and to those in power, usually governments, companies, or banks which are responsible for so much suffering. These letters appeal for precise action in a specific violation of a particular tribal people's rights.

Our members regularly hold demonstrations and vigils at the embassies of those countries where tribal peoples are dying, and these are often televised within the country they are aimed at. Apart from the obvious purpose of attracting public and press attention, we have noticed something else, quite unexpected, which has come from this. We have been told by diplomats inside the embassies that they have been genuinely moved to see our members—ordinary people, with no political axe to grind—wait outside on the street all night in cold winters, to uphold the lives of tribes they have never seen and will probably never visit. This selfless compassion has moved even hardened diplomats to reconsider their views, and embassy officials have spontaneously brought hot drinks and food out to demonstrators and told them how much they sympathize with our point of view.

We are running dozens of campaigns at any one time. Following a recent action targeting the Indonesian half of New Guinea, an American paper company abandoned a projected mill which would have wrecked the environment and destroyed the tribal peoples living there. We have seen projects for dam building shelved in Guyana and mining concessions withdrawn in Venezuela. We have seen the World Bank, which funds an enormous amount of destruction all over the world, pull out of dam projects in India and delay loans in Brazil because the Indians were not getting a fair deal. We have seen Indonesia cut the budget on its appalling transmigration programme where poor people are shipped to outlying islands and dumped on tribal lands. One of our most famous members, anthropologist Claude Lévi-Strauss, managed to stop a proposed car and boat rally in French Guyana which would have blitzed many Indian communities.

The Malaysian government has sent special trade delegations around the world to try to head off campaigns against the logging of tribal lands in Sarawak. They have asked to meet us for the express purpose of trying to persuade Survival to stop campaigning. Clearly, it works. Not always, and not always for long, of course. And some governments seem almost immune to pressure. In Bangladesh they are attacking and killing the hill tribes. We put pressure on the UN agency, the International Labour Organization, to send an investigative mission but the governments would not even allow them into the area.

Survival and the movement which it is catalysing and focusing is not a question of 'us', 'whites', 'Europeans', 'the North', (whatever term you like) trying to tell 'them', 'the Third World', 'the South', what they should do with their countries. The Brazilian pro-government press lapses into this accusation periodically, for example. It says that as Europeans have destroyed most of their own forests and natural resources, what right do they have to tell the Brazilians what they should be doing. But the point is that Survival is not a movement of the North. There are many supporters within the countries concerned. There are many Brazilians who support Indian rights. And there are even many Brazilians actually within government who support Indian rights.

This point is vital. It is very important that there is nothing in the least paternalistic about this struggle. This is not some kind of animal species that we are trying to save. We are not the developed world lecturing to the Third World, and we are not lecturing to tribal peoples either. This is a worldwide movement of concern.

Tribal peoples are well able to articulate their own defence if given the chance. And, of course, they do this better than any intermediary can. What we can provide more easily than they can is a knowledge of the outside, comparisons with other countries, analyses of international laws and forums and a general bringing together of widespread concern into a sharp focus—in brief, an efficient international organization with thousands behind it.

A few years ago the Guaymi Indians of Panama were threatened by the copper mine project of a UK-based multinational. We put the Indians in touch with Australian Aboriginals who had suffered at the hands of the same mining giant. They were able to explain how the company's promises of prosperity masked inevitable devastation. The Indians and Survival campaigned vigorously against it, and happily the mine was not built.

A couple of years ago we were instrumental in stopping the sale in the UK of the preserved head of a Maori man. We contacted Maori organizations in New Zealand who asked us to act. We went to court and succeeded not only in preventing the sale from taking place but in getting the auction house, one of the biggest in Europe, to agree never to deal in tribal peoples' dead bodies again. A small victory, but an important one which achieved worldwide press coverage. Many indigenous people are now struggling to have returned to them the bodies which are held in museum collections all over the world. Some believe that the afterlife of their ancestors is being severely disrupted until their bodies are disposed of in the traditional way. Imagine the pain of believing that your grandmother's soul has been prevented from going to paradise because her preserved body is displayed in some museum. This is not scientific enquiry, it is just barbaric.

Since its inception, Survival has put a great deal into supporting the growth of indigenous peoples' own organizations. Shortly after they formed, we were seeking funds for the new South American Indian federation, like the one in the Cauca area in Colombia which has done so much to regain Indian land from white landowners. They have paid a colossal price; on average one Indian leader murdered every month over the last fifteen years. Our Right Livelihood Award this year comes after several years of supporting indigenous organizations for the prize. Three years ago, the new Amazon Indian Confederation and its leader won one. Then last year it was the turn of the Sarawak native organization which has struggled so hard against the total logging of their lands.

We continue to offer our support to all genuine and representative tribal peoples' organizations. In our eyes the growth of this, their own movement, is the most important development we have seen in twenty years' work. We are not incurable romantics, green or otherwise, fighting a rearguard and doomed action against history. We are not talking about some inevitable drift into assimilation with white culture. Our experience of the last twenty years, the Aboriginals' experience of the last 200 years, or the Native Americans' experience of the last 500 years, shows us that is not actually what happens. Tribal peoples will struggle to hang on to what they have left and there's more tenacity in that struggle, more resilience in their cultures and ways of life than many think, or often would like to admit.

A few years ago I stayed with a group of Indians in Amazonia who had only just made peaceful contact with the outside world. A few months later I also visited the very first Christian mission to be established in the Amazon Basin nearly four hundred years ago. The mission is still there; it is quite a thriving Indian community with a rather progressive priest—himself an Indian from the highlands. Ten minutes walk away through the forest there were Indians living almost as traditionally as the uncontacted ones—they may have had transistor radios and clothes but their society and their self-sufficient economy was healthy and intact. Who is to say that in 400 years' time their descendants will not be there still?

Twenty years ago we heard many predictions that there would be no Indians left in Brazil by the end of the decade. These gloomy forecasts were wholly wrong. We are now optimists actually; not complacent of course, but hopeful that right thinking will prevail and the destruction of tribal peoples and their environments will stop. Tribal peoples will survive against extraordinary odds, but they do need the help of concerned people throughout the world. Survival seeks to bring this concern together and forge it into an effective weapon to put into the hands of the people themselves.

Our goal is to build a strong and well focused worldwide movement to help tribal peoples. We cannot stress too much that Survival is its members' work. The Right Livelihood Award will be an enormous help to them in this task which is desperately needed now more than ever. And it is on their behalf, as well as on behalf of all threatened tribal peoples, that I thank the Foundation for awarding us this prize.

Statement of the Yanomami People: To All the People of the Earth

DAVI KOPENAWA YANOMAMI

THE GOVERNMENT ISN'T RESPECTING US. It thinks of us as animals. We have the right to defend our rights. There are many people who help us, but if we don't do anything, they can't help. If we send a letter to the government, these people will put pressure on the government, and take other actions. I have many things to say that I've been thinking about. I am a Yanomami.

We Yanomami thought that the white man was good for us. Now we see that this is the final invasion of indigenous lands; all others have already been invaded. They came to take away our land. They are taking it away. There are people from the outside who are instructing the Brazilians to destroy our lands. The same thing happened in other places, with our Indian brothers in North America. Now it's happening here on our land. The government shouldn't do this, for the government knows that we are the first Brazilians, that we were born here, that we are called Yanomami.

Our name is known throughout the world. We don't use money, shoes, clothes, and few Yanomami understand what is happening. The government took us by surprise. Now I am beginning to understand. The government doesn't know our ways, and our way of thinking. We also don't know the ways and way of thinking of the government. The government only understands this business of money. Our way of thinking is based on the land. Our interest is in preserving the earth, so as not to create sickness for the people of Brazil, not only for the Indians. The gold-panners, the squatters don't have land, so they invade the land of the Indians. If they had their own lands, they wouldn't invade our area.

I also see how the whites suffer in the cities from hunger, from the high prices, from the lack of housing, the lack of food. All of them are suffering. They are concerned, but don't have the courage to complain to our leader, the President. He is also deceiving the people because he also takes orders from other countries to destroy our lands, build highways. The government also takes orders from other wealthy men—it asks for money from them and they give it to the government to abuse our lands. The rivers, the fish, the forests are crying for help, but the government doesn't know how to listen. It says that we will die of hunger if they close the gold mines. But I say that if they don't stop gold-panning, then we will die of hunger. If the government stops the gold-panning, we will plant sweet potatoes, bananas, yams, taro, papaya, sugar cane, pupunha fruit, and then no one will die of hunger. We Yanomami want to keep our land. We don't want to stop our traditional ways of life. We haven't yet lost our language nor our land, which is why we are struggling today. This government is our chief, but it isn't helping the Brazilian people live in peace.

Today, all the Indians of Brazil are united, and we don't want to fight with other kinsmen. They are already beginning to deceive us, as happened with the Macuxi and other Indians. They are saying that the priests are no good, that they don't give us presents, that Davi is no good. They set brother against brother in order to weaken us. The other Yanomami chiefs who have never come to Boa Vista, who have never had contact with the whites, don't know what's happening here. I know, and they want to use me because I am more known, but I won't let them use me. There, they're using the weakest first. Those chiefs who don't speak Portuguese think that the gold-panners give food and clothes, but then time passes and the whites begin to say that the Indians aren't worth anything, that they don't work, that they only beg. And they will call us Urubutheri (vultures), who don't hunt anymore, who don't fish anymore, who live off what's left over, what's left over from the white man's plate. They say that we don't know how to work or to fish anymore, and that we only beg. They say that we forget how to gather fruits in the forests, forget our ways and our language.

I don't want to lose that, and I don't want to let gold-panners into my community. I want us to stay as we were before. But we are suffering today. I always remember our grandchildren; they will suffer more than we are suffering if we don't fight to defend and save the lives of our people. The government says that the land isn't ours; it says that we can fish, make gardens, hunt and use the lakes and rivers. It says that we are using the government's land, but the land is ours. This we have known for many years. The government isn't good to us. We fight and fight but it doesn't give us what we ask for. That's why it's so difficult to get our lands demarcated, because the government doesn't want to demarcate Yanomami lands. There are many things the government wants to use on Yanomami lands—there are minerals, there's gold, cassiterite, timber, and the land can be cultivated.

And the President only speaks to us hidden in his office. He doesn't call us to decide or to know if the Indian agrees with him or not. But I'm not fooling around in this struggle. I am here to defend my Yanomami people. And not just my people, but also the Wapixana, Ingariko, Macuxi, and other kinsmen. We are trying to help. We can give help to those who don't know how to defend themselves. We can explain what is happening to those who don't understand. We Yanomami are dying from diseases—malaria, flu, dysentery, venereal diseases, measles, chicken pox—and other sicknesses that the Indians never knew before and that were brought in by the gold-panners from outside. These sicknesses we can't cure; shamans can't cure them. Gunshot wounds, shamans also can't cure. Sicknesses of the Indian, shamans can cure, but sicknesses of the whites, we can't cure.

I've always asked Funai's help [Funai: the government agency for Indians], but Funai doesn't take the necessary measures; I've also asked President José Sarney to remove the gold-panners from Yanomami land. The President just lets them invade more. We Yanomami think that he doesn't like to help the indigenous people of Brazil. I know that he is against us. He doesn't want to demarcate our land. I have had a lot of news from my kin who live at the headwaters of the Catrimani River, the Mucajai River, and the Palimiu river, on the border of Venezuela. My kin told me of four empty malocas [houses of the Yanomami], everyone had died. Children, adults, and young men. In the Xideatheri, Ahisahipiktheri and Pahaiatheri malocas, everyone dies and others go on dying because of the lack of assistance. Funai knows that many Yanomami are dying, but it's doing nothing. There are only a few people in Funai who want to work, but they don't get support. In other communities of the Mucajai River, my kin are suffering because of the gold-panners, who have gotten them used to drinking cachaca mixed with caxiri [a mildly fermented beverage], and now they have become sick and don't know what to do. And also there is a lot of venereal disease and malaria.

On the Catrimani River and on the hills of the Lobo D'Almada River, the gold-panners are making their houses. They have built airstrips, made gardens, and now they want to make a town. This will be very dangerous for the Yanomami. I know that if they make a town there, it will be bad for the Yanomami because they will get sick. Our kin of Opiktheri are being deceived by Zeca Diabo, a gold-mining businessman. He is trying to set up a fight among the Indians. Zeca Diabo helps in giving the Indians food and clothes, in teaching them how to work, to make gardens, to plant and gather rice so that the Yanomami then get used to working for him. But I don't think it's good that the whites teach the Indians how to work. We Yanomami already know how to work; for many years we have known how to plant, to clear the forest and to cut underbrush for gardens. The Yanomami don't die of hunger; they only die of diseases. They have everything to survive where there are no gold-panners. My kin don't beg for food from the whites; they only ask for food where there is a gold mine which usually does away with everything. We already raise animals, we already have tapirs, peccary, curassow, wild boar. We plant banana; we have everything in our forest. There is no need to teach us how to work in the white man's way. The white man's way is very difficult for the Yanomami. Our way is better than the whites' because we preserve the rivers, the streams, the lakes, mountains, game animals, fish, fruits, acai, bacaba, castanha, cacao, inga, buriti, what's already there, what Omam created. I, Davi Kopenawa Yanomami, want to preserve it all. The white man has no respect for nature; he doesn't know that it is good for him. He has to learn from us.

The government has cut up our land, divided it into little pieces. The National Forests are our land; the 'island' reserves [nineteen mini-reserves delimited by the government] are not good for anything, only to trick the Indians, to leave them stuck like pigs in a pen. We Yanomami want a single and continuous area for our people in order to be able to live in peace, without having to fight the government, the military, the gold-panners, or anyone else. The Yanomami don't seek to invade other people's lands. The Yanomami respect the white man's lands.

At the Serra do Surucucu, more Yanomami were killed in August. Funai has done nothing against the killings. The police have never taken these criminals prisoner. We Yanomami are dissatisfied; we are revolted with Funai and the government because the government doesn't want to find a solution to our problem, the problem of Yanomami land. We have many hills on our lands; Koimik is the Pico da Neblina. Hakomak is the Peito da Moca. Watorik is the Pico Rondon. Kuumak is the Serra do Taraqua. Yapihukak is the Serra do Lobo D'Almada. Arahaikyk is the Serra do Catrimani, and there are many others. The spirits of nature, the Xapori and Hekura, live in these hills. Between the hills there are the Xapor trails; no one sees them; only the shamans know these connections. The hills are sacred places, places where the first Yanomami were born, where their ashes have been buried. Our elders left their spirits in these places. We Yanomami want these hills to be respected. We don't want them to be destroyed. We want these places to be kept as they are so as not to put an end to our past and our spirits. We invoke the Hekura to cure our sick. For many years we have used them; there is no end to them. Omam left these spirits to defend the Yanomami people. Omam is very important for the Yanomami Indians. He gave rise to all of them, to the whole world. So it is very important to keep the hills where his spirit lives as they are. I would like the whites to understand this sacred history and to respect it. We Yanomami want to see the whites on the side of the Indians in not letting our lands be invaded. We want the whites to help in defending our land so as to protect our lives. I, Davi Kopenawa Yanomami, want to help the whites learn with us how to make a better world.

4

People's Knowledge:

The Health and Development of Community

INTRODUCTION TO THE PROJECTS

Rosalie Bertell
1986
'for raising public awareness about the destruction of the biosphere and human gene pool, especially by low-level radiation'

DR ROSALIE BERTELL was born in 1929 and received her Doctorate in Biometrics in 1966 at the Catholic University of America. She has been working in the field of environmental health since 1970. Bertell was instrumental in founding the Ministry of Concern for Public Health in Buffalo, N.Y. in 1978. At present, she is President of the International Institute of Concern for Public Health in Toronto, Canada, and is one of the founding Commissioners of the International Medical Commission based in Geneva, which will work with health professionals to implement the concept of health as a human right. Bertell is also Editor in Chief of *International Perspectives in Public Health*, and author of *No Immediate Danger: Prognosis for a Radioactive Earth* (Womens' Press, London, 1985).

Dr Rosalie Bertell
International Institute of Concern for Public Health
830 Bathurst Street
Toronto, Ontario M5R 3G1
Canada

Alice Stewart
1986
'for bringing to light in the face of official opposition the real dangers of low-level radiation'

DR ALICE STEWART was born in 1906 and had a distinguished early career as a clinical physician, becoming in 1946 the youngest woman to be elected a Fellow of the Royal College of Physicians. This was also the year in which she turned to Social Medicine, joining the Unit of that name in Oxford. In 1955 she and her colleagues noticed the rapid increase in leukaemia among children, which seemed likely to have environmental causes, and the idea for what became the Oxford Childhood Cancer Survey, or just Oxford Survey, was born.

One of the key early findings of this survey (1958) was that children who died of leukaemia or cancer had been X-rayed in utero twice as often as healthy children. This controversial finding led eventually to the cessation of X-rays for pregnant women and confirmed Stewart's interest in and focus on the health effects of low-level radiation. The Oxford Survey was extended to adult cancer sufferers and further data was collected which supported the original conclusions, which were finally accepted by the International Commission for Radiation Protection (ICRP).

> Dr Alice Stewart
> Department of Social Medicine
> University of Birmingham
> The Medical School
> Edgbaston, Birmingham B15 2T, UK

Dr Aklilu Lemma & Dr Legesse Wolde-Yohannes
1989
'for their pursuit of knowledge about the molluscicidal and other properties of the endod plant and their twenty-year perseverance in working to overcome the biases against Third World science in the Western medical establishment'

For twenty-five years two Ethiopian scientists, DR AKLILU LEMMA and DR LEGESSE WOLDE-YOHANNES, have been struggling to win international recognition and support for their discovery of a cheap, community-based prevention of a chronic Third World disease. The disease is bilharzia, or schistosomiasis, a debilitating and eventually fatal illness caused by flatworm infestation of the liver and other organs, which afflicts more than two hundred million people in seventy-four countries of Africa, Asia and Latin America.

Present molluscicides, to kill the snail-carriers of the disease, and therapies for bilharzia are far too expensive for the communities that need them. Lemma's discovery, in 1964, was that the fruit of a common African plant, the endod or soapberry, acts as such a molluscicide. The endod berry has been used for centuries by African women as a soap to wash their clothes. Lemma, who took a doctorate from Johns Hopkins University in the United States, made his discovery by observing a pattern of dead snails in a river downstream from where some women were doing their washing.

In 1966 Lemma established the Institute of Pathobiology in Addis Ababa University, and for the next ten years directed a team systematically to research endod. He was joined in this work in 1974 by Wolde-Yohannes, who received his doctorate from Technical University in Hannover, West Germany, and continues with endod research at the Institute to this day. Lemma has established the Endod Foundation in Ethiopia as the holding body of the various patents and processes that were developed.

Given the immense promise of the discovery, confirmed by the early research, progress in making this molluscicide available to the people who so desperately need it has been tragically slow. One of the reasons for this is that Third World research is still not taken seriously in the West and does not qualify products for international acceptance. International organizations will not promote a product without expensive Western research, and most Third World governments cannot or will not introduce a new remedy without such support. Moreover, endod's very cheapness and simplicity mean that its commercial possibilities are not attractive enough for private enterprise to invest the necessary resources to win toxicological clearance.

In the last few years, Lemma's and Wolde-Yohannes' persistence seems to have made progress towards overcoming these problems. The support of key scientists and donors in the West has opened the doors to the necessary laboratory and field trials. Endod now offers the prospect of being a poor country's solution to a poor person's disease.

Dr Aklilu Lemma
Deputy Director
International Child Development Centre
Piazza S. S. Annunziata 12
50122 Firenze, Italy
Tel: 39 55 2345258

Dr Legesse Wolde-Yohannes
Institute of Pathobiology
Addis Ababa University
PO Box 1176
Addis Ababa, Ethiopia
Tel: 251 1 135728

John F. Charlewood Turner
1988
'for championing the rights of people to build, manage and sustain their own shelter and communities'

JOHN F. C. TURNER, an Englishman, was born in 1927. Since 1957 he has been involved with the practice and developing of the theory and tools for self-managed home and neighbourhood building in Peru, the United States and in the United Kingdom. Turner graduated in architecture from the Architectural Association in London in 1954 and worked in Peru from 1957 to 1965, mainly on the advocacy and design of community action and self-help programmes in villages and urban squatter settlements. From 1965 to 1967, he was a Research Associate in Cambridge, USA, at the Joint Centre for Urban Studies of the Massachusetts Institute of Technology (MIT) and Harvard University and then lectured at MIT until 1973. Returning to London, he was a lecturer at the Architectural Association and the Development Planning Unit, University College London, until 1983, when he resigned to devote himself full-time to his non-profit consultancy AHAS, which he established in 1978 with his wife and Co-Director, Bertha, and two other colleagues.

AHAS
51 St. Mary's Terrace
West Hill, Hastings
East Sussex TN34 3LR, UK

Seikatsu Club Consumers' Co-operative
1989
HONORARY AWARD
'for being the most successful model of production and consumption in the industrialized world, aiming to change society by promoting self-managed and less wasteful lifestyles'

The Seikatsu Club Consumers' Co-operative (SCCC) is a unique consumers' co-operative in Japan. With 170,000 family members, the organization combines formidable business and professional skills with strict social and ecological principles and a vision of a community- and people-centred economy which provides a radical alternative to both socialist and capitalist industrialism.

SCCC traces its foundation back to 1965 when a single Tokyo housewife organized two hundred women to buy three hundred bottles of milk to reduce their price. It has now developed into a significant business enterprise. In 1987 its turnover was forty one billion yen and it employed seven hundred people. The business centres on the distribution to 'hans', local units averaging eight members, of four hundred different products, of which sixty are original brands and sixty per cent by value are primary products, like rice, milk, chicken, eggs, fish and vegetables. The Club will not handle products which are detrimental to members' health or the environment. It currently owns two organic milk factories and manufactures several varieties of soap, including one from recycled cooking oil.

To supplement the 'hans', the SCCC set up women's workers' collectives to undertake both distribution and other service enterprises, including recycling, health, education, food preparation and child care. By December 1987, fifty-seven such collectives were employing 1,550 SCCC members. The Club has also established a non-profit insurance company for members.

In their campaigns against synthetic detergents, Club members realized the importance of the political process and formed independent Networks in different prefectures to contest local elections. In 1979 the first Network member was elected to Tokyo city government and there are now 33 Network councillors in Chiba, Tokyo and Yokohama, all of whom are women. The manifestos of these Networks, now numbering twenty-two, resemble European Green manifestos: very environmental, peace-oriented and anti-nuclear, emphasizing local participatory democracy and equal status for women.

Seikatsu Club Consumers' Co-operative
2-26-17 Miyasaka
Setagaya-ku
Tokyo, Japan
Tel: 81 3 7060031

Nucleogenic Illness: The Work of the International Institute of Concern for Public Health

ROSALIE BERTELL

December 1986

IT IS A DEEP PLEASURE FOR ME to gratefully accept the Right Livelihood Award for 1986 on behalf of myself and those who have worked with me at the International Institute of Concern for Public Health. Sweden is gaining an international reputation for its extraordinary efforts on behalf of global justice and peace, and for its yearly search of the global community for creative and concerned persons and organizations which could use some encouragement and financial assistance. This is a much valued service to the forming global village. In the long run it will, I think, be more humanly productive than increased airport security, military exercises, nuclear threats and development of crowd-control technology. This contrast between a system of encouragement and co-operation, on the one hand, and a system of threats and forcible control, on the other, lies at the centre of the global crisis. It poses a clear choice for the future, on which will depend the survival or disintegration of civilization.

I have been told that the Vikings travelled both to North America and to Russia, demonstrating a desire to discover new lands and peoples, as well as to bridge global differences. While travelling in Sweden, I noticed also that the women of these Viking explorers managed large farms and produced food, shelter and clothing for the family. Swedish development relied equally on establishing security and co-operation at home, and expanding consciousness and concern through adventurous sallies forth to foreign and sometimes very distant lands. The recognition of the work of the International Institute of Concern for Public Health by the Right Livelihood

Foundation focuses global attention on the necessity of developing security for the global village, meeting its need for clean air, water, food and a healthy habitat, as well as fostering clarity of vision on co-operation and development. These provide an essential balance to the high technology atmosphere of First World research as it sallies forth into outer space and sub-microscopic life.

Since Chernobyl, the fragility of human health, our food supply and the threat to domestic tranquillity in the face of lethal nuclear fallout has become apparent to all. Yet less than five per cent of the poisons contained at the Chernobyl reactor escaped into the environment. The remainder has been buried under a cement mausoleum, in spite of the failed US attempt to do the same at Runit Island, Enewetok Atoll in the Marshall Islands. Forty years into the Nuclear Age, the Chernobyl accident found nations in panic and disarray, with conflicting public health and radiation protection criteria, empty assurances for the public at risk, and questionable practices of disposal of contaminated food. Milk was dumped on farm and pasture land; workers were sent out to harvest contaminated crops without proper clothing or respirators; and in Sweden military exercises were planned for a region of the country which had experienced the highest nuclear fallout.

Radiation-induced cancers resulting from Chernobyl and the 1,600-plus nuclear weapons explosions or accidents which have already occurred on this planet constitute only a small fraction of the tragedies caused. There will be embryonic, foetal and infant deaths, congenital diseases and malformations, various degrees of genetic damage to future generations, and a variety of human tragedies officially lumped together under the phrase 'ill health' in radiation protection literature. Most of these will not be officially recognized as problems in a world grown callous about random murder by technology.

A report, soon to be published from the US Council on Economic Priorities based on vital statistics has conservatively calculated that there are 9,000 excess deaths a year in the US attributable to the routine operation of commercial nuclear reactors. I was able to document 100 excess deaths of low birth weight infants born downwind of normally operating 'state of the art' nuclear reactors in their first five years of operation in Wisconsin. These are only small glimpses at the tragic story, since many who are casualties do not die. Let me read from the sworn affidavit from a woman who was a non-fatal casualty at Three Mile Island during the time of the serious nuclear reactor accident in 1979:

> I provided the following information to Jane Lee and Majorie Aamodt on May 11, 1984.
>
> On Friday evening, March 30, 1979, I was standing on the front porch of my home. My home faces south. It was raining, and the wind was blowing. All of a sudden the cat that had been let out began to howl in a most unusual way. I had never heard a sound like that from this or any other cat. I called the cat by name, however it did not come home . . . I went over to the bannister and leaned over to call the cat again. . . . Suddenly, the wind stopped; there was a movement in the limbs of the trees next to the porch, and a wave of heat engulfed me. The gust of heat brought the rain over me. Then the wind started again. This all happened in about one minute. I was so startled that I went in, taking the cat, who had by now come up on the porch. I wiped my arms and legs with the towel. I noticed that my arms and face were pink. I applied a lotion because my skin felt tingly.
>
> On Saturday morning, my skin was a darker pink, and there was an itch at the front of my scalp. This was the only part of my scalp that was not covered by a scarf. When I went to church on Sunday, my friend commented that I looked healthy and sunburned. On this day, hard little lumps, a little bigger than a pinhead appeared on my forehead and into the hairline. . . . About three weeks later, I noticed that a lot of grey hairs had appeared across the front of my hair. When I washed my hair that week, my comb was full of hair.
>
> . . . In the subsequent weeks, the skin on my forearms and neck turned darker and was scaly. This condition lasted for several years. There is some permanent discoloration, however it is not prominent. My forearms were, and continue to be, very sensitive to the sun, becoming itchy with exposure. I try to avoid sunlight. I have also noticed that if my arms are injured, the bruise will

last longer than was normal for me prior to the event described above.

Of greater concern to me presently is the loss of the function of a kidney. Toward the end of November 1983, I was in renal failure. My doctor described my condition as an unusual case. He stated that one of my kidneys had died. I was in Holy Spirit Hospital under the care of two doctors. I have not fully recovered, and I have not been able to resume my customary social and household activities.

I live on a farm with my husband. We were not able to evacuate during the accident, although I wanted to leave, because my husband would not ask anyone else to stay to do his job of caring for the animals. Despite our continual attention to the cattle, we experienced the first deformed calves ever born on our farm the following spring. The calves' heads hung to one side until they were six months old. Their necks appeared twisted. I also noted that the Norway maple by our home had deformed leaves which were curled at the edges.

No doubt many similar stories could be told in the Soviet Union, in Sweden, and Europe since Chernobyl. With no biological testing of victims it is hard to prove causality. None of these problems are considered serious in the eyes of nuclear experts because they have not yet caused radiation-induced cancer deaths. We need a new word to describe this random damaging of life. I would suggest using 'nucleogenic' or 'technogenic' illness to describe these induced abnormalities.

In the face of such lack of sensitivity to human health considerations, concern for the future viability of the human race, placed under great stress by militarism and high tech, is rational, not irrational. Human passivity has permitted between 10 and 20 million deaths or serious casualties due to the nuclear fuel or weapon cycle since 1945, not including the tragedies of Hiroshima and Nagasaki. These conservative estimates do not include deaths and cripplings especially among indigenous people and in the developing world due to the financial burden of the arms race and warped First World economic priorities.

Careless damage of the biosphere compounds the problems, as people physically less able to cope than their parents and grandparents will try to live in an increasingly hazardous environment. The increasing hazard stems from the intractable problems of uranium, nuclear, and other military pollution and waste. The raised voices of the exploited at this moment in history are indicative of the deepest and best survival instinct.

With these global problems in mind, I would like to make some suggestions for action by the people of the United Nations in behalf of future living on this lovely habitable planet:

1. Co-operation and participation in international meetings and conferences dealing with the plight of radiation victims (for example the first European Conference on the Medical Management of Radiation Recipients held in Amsterdam in 1987, and the first international conference on this topic held in New York also in 1987). The Institute believes that visibility of the radiation victims and alerting the medical community to their problems is essential for motivating changes in governmental behaviour. Those most seriously injured by national ambitions and rivalry need a platform for making known the severe negative impacts of such policies before the eruption of nuclear megadeath.

    At such meetings there will be a forum for physicians, scientists, lawyers and others intent on healing the incredible nucleogenic and technogenic sickness with which we abuse ourselves, our brothers and our sisters in the name of progress and national security. Only 'common security', a phrase attributed to the late honoured Swedish Prime Minister, Olaf Palme, is acceptable in a global village.

2. Begin to dismantle the pseudo-scientific establishment which has rationalized nuclear addiction with self-appointed and self-perpetuated global advisory bodies. The International Commission on Radiological Protection needs to be dismantled first and replaced by an International Institute for the Safeguarding of Communities and Workers from Preventable Exposure to Ionizing Radiation. Unlike its predecessor the ICRP, this new institute

should be appointed by relevant scientific and public health organizations, be subject to scientific peer review, be responsible for public health recommendations rather than risk/benefit tradeoffs, and be composed of persons trained in epidemiology, public health, toxicology, and worker health problems rather than of radiation users.

3. The International Atomic Energy Agency needs to be relieved of its mandate to promote nuclear energy. It might be retained temporarily to assist in the dismantling of this industry globally. However the monitoring of health and safety relative to this industry needs to be placed within a new agency not mandated to promote nuclear technology. An international energy agency should be established and mandated to implement the resolutions of the United Nations Conference on New and Renewable Sources of Energy held in Nairobi in 1981. Unbiased scientific data on non-nuclear energy technology must be quickly made available to developing nations. High pressure selling of unwanted First World nuclear technology to already exploited developing countries will exacerbate present health, environmental and financial difficulties.

4. Begin phasing out the United Nations Security Council which merely reinforces the dominant position assumed by the nuclear nations in the face of international disorder. If we truly want to assume stature as a mature global family of nations we need to extend responsibility to the people of the world through a representative assembly, rather than assign abnormal political power to countries with weapons of mass destruction. We need to invest power in international treaties, conflict resolution mechanisms and the World Court.

5. Demand the cancellation of the United Nations Conference on the Peaceful Uses of Nuclear Energy (PUNE) scheduled for 1987. High technology, not wanted in the developed world, should not be pushed on the developing world especially under United Nations auspices. Food irradiation will be a major issue at the PUNE conference. Inconclusive research done on well-nourished laboratory animals is being used to substantiate promises of healthful food to populations already severely decimated by

hunger, malnutrition and starvation. Food irradiation delivers stale food disguised as nutritious, polluted with unique radiolytic by-products, to children suffering from deprivations which are already life threatening. Published research from India indicates that malnourished children fed irradiated wheat developed polypoidy, a cellular phenomenon also observed in cancer, during viral infections, in senility, and after radiation exposure. Long term health implications, problems of transportation of irradiation sources, worker health and safety, and disposal of the radioactive waste from this new industry are poorly addressed, if at all, by the proponents of food irradiation.

It is my sincere hope that the awarding of the Right Livelihood Award to Dr Alice Stewart, and to myself, will mark a close to the era of global nuclear expansion and deception with respect to the hazards of ionizing radiation. I hope and pray that it will be the beginning of reality therapy, of healing the pre-World War III radiation victims, of the dawn of an era of serious efforts to consciously build the infra-structure of the forming global village. Domestic or common security is still the 'sine qua non' of human survival and development. It can never be sacrificed for adventure, profit or political gain.

The people of the global village are longing to share in the benefits of human endeavour, not the garbage; to share in the wholesomeness of life and not to be handed death. It is my heartfelt wish that the good begun here will flow forth as a river of life, justice and hope for those people most broken and needy in our global village. May it mark a firm choice for the Global Encouragement and Co-operation System initiated in Sweden, and an end to the Threat and Forcible Control System which has been a global plague for centuries and culminates in worldwide hostage-holding to nuclear terror.

The Oxford Survey of Childhood Cancers

ALICE STEWART

December 1986

A LENGTHY ASSOCIATION with the Oxford Survey is clearly the reason why I have been honoured with one of the 1986 awards of the Right Livelihood Foundation. This Foundation is anxious to see established 'a science of permanence' in relation to many of today's problems. Therefore, this occasion would seem to be an appropriate one for describing a survey which, whatever its other failings, could never be accused of taking too hasty a look at the problem of early cancer deaths. The name is a reminder that for twenty years the survey was an Oxford University activity, but for the last twelve years the headquarters have been in Birmingham University. Like the tortoise in the fable, the survey was never designed for sprinting. However, again like the tortoise, it has shown great endurance and an ability to outclass rival projects which once seemed certain of winning all the prizes. The credit for these achievements belongs to many, many persons, but it gives me great pleasure to acknowledge my special indebtedness to David Hewitt and George Kneale, without whom there would have been no story to tell.

ANTECEDENTS

The first English university to promote social medicine to full academic status (by establishing a Chair in the subject) was Oxford. This was a wartime move prompted by a Nuffield benefaction which was intended to keep a research and teaching institute in funds for ten years. It led, in 1942, to a distinguished London physician, John Ryle, coming to Oxford as the first Professor of Social Medicine and accepting the following brief: (1) to investigate the influence of social and genetic factors on the prevalence of human disease and disability; (2) to find both new ways of identifying factors which interfere with full development and maintenance of health and new ways of protecting society from the effects of these factors; and (3) to provide for the teaching of the new subject to clinical students and postgraduates.

The early years of the new institute were uneventful but by the time of Professor Ryle's death, in 1951, both the Nuffield benefaction and the initial (wartime) enthusiasm for the subject were exhausted. Therefore, on the advice of a science faculty which naturally regarded experiments as a better research investment than surveys, the university disbanded the institute, leaving in its place only one person with no clearly defined duties except 'to do such teaching as the clinical professors require'.

Overnight I had become 'University Reader in Social Medicine' with nothing to do and no special training in a subject which clearly needed statistical as well as medical expertise. All I had was considerable experience in clinical medicine and a firm belief that it only needed a few completed surveys for the university to recognize its mistake and resume full support of the subject.

During the war, while working in the Nuffield Department of Clinical Medicine, it had occurred to me that a survey might be the best way of resolving one of the problems referred to the department by the Medical Research Council (MRC). It was a typical wartime problem (requiring quick action) and my rapidly improvised survey was so successful that similar activities became a regular accompaniment of my clinical work. It was this wartime experience which led to my being appointed as first assistant to the Professor of Social Medicine in 1946.

After crossing the Rubicon between clinical and social medicine, my first concern was a survey of workers in shoemaking factories. The original purpose of this research project was to discover how tuberculosis is spread between fellow workers but, like all good surveys, it had branched out in various directions. Therefore, when the institute was disbanded, I still had helping me an erstwhile Medical Officer of Health, Josephine Webb, and a young statistician, David Hewitt, both of whom were paid for by an MRC grant. We were anxious to remain as a research team but in order to do so we had to find a new project and new funding.

These events were taking place at a time when, both in Europe and in North America, deaths from leukaemia were increasing at an alarming pace. During the next decade the rest of the world also exchanged low, prewar levels of leukaemia mortality for the much higher (though stable) rates of today. But in 1951 there was considerable anxiety lest we were on the brink of a leukaemia epidemic.

In spite of the rising death rate, leukaemia was still a rare disease and, except in Hiroshima and Nagasaki, there were no local epidemics. Therefore, a survey to discover the causes of the rising death rate was not regarded as a practical proposition. However, a close scrutiny of several sets of national statistics, by David Hewitt, had revealed something which made us think that perhaps one facet of the rising death rate might be identified by interviewing mothers of children who had recently died from leukaemia and comparing what they had to say with similar information from mothers of live children and children who had died from other malignant diseases.

What David had done was to show that, instead of a few leukaemia deaths being evenly distributed between infants and older children (which was the prewar pattern), there was now a conspicuous peak of mortality between two and four years caused, not by the type of leukaemia which was threatening adults (myeloid), but by the opposite type (lymphatic). Therefore, it was possible that young children were the main sufferers from an antenatal event which had escaped the notice of obstetricians but might be retrieved by systematic questioning of mothers.

We should have realized that the Medical Research Council was unlikely to be impressed by a research project which depended upon memories of lay informants and would certainly require several data collecting centres. However, although we failed to obtain an MRC grant, we were given £1,000 by the Lady Tata Memorial Fund for Leukaemia Research. There were no conditions attached to this windfall and I was free to do what I liked. Furthermore, the university had finally agreed to my having a statistical assistant, and Dr

Webb was told that she could remain on the external staff of the MRC indefinitely. Therefore, we decided to do two things: obtain the death certificates of all children who had died from leukaemia and other forms of cancer in recent years (i.e. 1953-55), and seek the help of Public Health Departments. These departments had an ideal source of live controls (in their registers of live births) and might agree to do paired interviewing of cases and controls since there were so few cancer deaths of children.

ACHIEVEMENTS

We eventually persuaded all County and County Borough Health Departments to serve as data collecting centres and gradually set in motion a survey which not only became coextensive with the whole country (i.e. with England, Scotland and Wales) but was so successful that, in no time at all, over eighty per cent of the 1953-55 cancer deaths were traced and matched with live controls. We even obtained early warning of the prenatal X-ray finding (by detecting a significant case/control difference in the first batch of completed interviews). Therefore, from the onset there was systematic checking of mothers' claims against actual records of obstetric X-rays with foetal involvement. But although we had achieved one of our goals (by finding an antenatal event with cancer associations) we soon discovered that we still had no explanation of the early peak of leukaemia mortality.

You will recollect that the original reason for the survey was to discover why young children (but not infants) were being increasingly threatened by a certain type of leukaemia. What we had found was evidence that an uncommon event (only seven per cent of our controls had records of foetal irradiation) had probably caused a few deaths which were scattered among leukaemias and other malignant diseases and insufficient to leave any mark on vital statistics. Therefore, if we had not accidentally found these radiogenic cases there would have been no incentive for later searches of cancer effects of prenatal X-rays and other sources of low level radiation.

The necessarily small numbers of deaths caused by a few children being briefly exposed to a small dose of X-rays shortly before birth (and the even spread of these cases between leukaemia and other malignant diseases) were to cause us many future headaches. But, in 1958, the main problem was to convince sceptical radiobiologists that there had not been biased reporting of X-rays by mothers of live and dead children. This explanation of our finding was eventually accepted by all radiation protection committees after they decided that neither animal experiments, nor screening of A-bomb survivors for cancer deaths, nor similarly screening of X-rayed children (prospective surveys) produced any evidence of a cancer risk from foetal irradiation. Therefore, for many years faulty design of survey remained the 'official' reason for our prenatal X-ray finding and also a reason for giving no encouragement to case/control or retrospective surveys.

According to this assessment of OSCC data, there was no reason why the X-rayed and non-X-rayed cases should have different ages. However, short of both groups consisting of cases which had been initiated during the third trimester of foetal life (i.e. during the usual period for practising obstetric radiography) one would not expect the radiogenic and spontaneous cases to be affecting exactly the same age groups. Which group would prove to be the younger one would depend upon whether the spontaneous cases were initiated before or after the third trimester. Therefore, if we could only establish a significant difference between the ages of our X-rayed and non-X-rayed cases we could (a) confirm the existence of radiogenic cases without reference to our live controls and (b) discover whether to look for other causes of juvenile cancers among prenatal or postnatal events.

This was the reasoning behind a decision to revive the Oxford Survey and put it on an ongoing basis by continually adding to the original series of 1953-55 case/control pairs. Provided data collection continued until 1971—by which time all 1953-55 births would have merged with the adult population—we would have the equivalent of an enormous prospective survey (i.e. a fifteen-year follow-up of more than two million live births). This would provide the correct basis for comparing ages of X-rayed and non-X-rayed cases and sufficient numbers of X-rayed cases for studying several types of cancer.

This was our main objective when we again approached the Public Health Departments and again obtained their voluntary co-operation. But we also realized that there would be other rewards. The interviews with case and control mothers had covered many topics including maternal age, birth rank and social class; also pregnancy illnesses, drugs and X-rays; postnatal infections, inoculations and X-rays, parent's occupations, family histories of cancer and, eventually, the ultrasound replacement of prenatal X-rays. Therefore, by continuously extending the data collection period we would eventually have the necessary records for testing several hypotheses. What we did not anticipate—though it actually happened—was that during the data collection period new methods of statistical analysis would be invented which exactly suited the needs of any large data base consisting of cases with matched records. These new techniques not only enhanced the value of OSCC data but also restored faith in retrospective surveys (i.e. surveys which often went unfunded because they had less in common with experiments than prospective surveys).

Shortly after the restart of data collection we learnt that a prospective survey of three quarters of a million live births in the North Eastern states of the US had found both a raised cancer death rate for X-rayed children and a significant difference in the ages of X-rayed and non-X-rayed cases. These findings gave us great encouragement but did not alter our plans. We were reasonably certain that the new survey design was foolproof and would eventually achieve far more than mere confirmation of the X-ray hazard. For example we already anticipated being able to test three hypotheses advanced by American scientists: first, that preconception as well as prenatal X-rays were causes of juvenile cancers; secondly, that the association between prenatal X-rays and cancers was caused not by the radiation but by the medical reasons for the X-rays and, thirdly, that the radiogenic cases were confined to children who had an inborn susceptibility to radiation, infections and allergies. Eventually, none of these theories was able to withstand tests based on OSCC data. But this is to anticipate the analytical work of George Kneale.

George was still a schoolboy when the Oxford Survey began. But he began working on the project even before he had added an

THE OXFORD SURVEY OF CHILDHOOD CANCERS

Oxford University diploma in Statistics to his other qualifications (i.e. an honours degree in Chemistry). He actually started work in 1962 and, ever since then, has been adding to a repertoire of new ways of analysing any survey data concerned with cancer effects of radiation or other factors with cancer associations. Even before we moved from Oxford to Birmingham University (in 1974) George had greatly enhanced the value of OSCC data by inventing a new method of analysing truncated contingency tables. This was the form necessarily taken by OSCC data when testing for cancer effects of prenatal events. With the new method we could make almost as much use of remote and recent birth cohorts as of the cohorts which could be followed from birth to fifteen years of age. Therefore proof that radiogenic cases existed (and were a fraction older than spontaneous cases) was obtained well in advance of the expected date. Armed with this knowledge we were prepared for two eventualities: first, that all the OSCC cases might have moved from a safe in utero environment to a hazardous postnatal environment after cancer induction; secondly, that for most of the non-X-rayed cases this move might have coincided with a later stage of the cancer process than for most of the X-rayed cases. In short, we had learnt both the extreme importance of discovering whether there were changed reactions to other illnesses when they coincide with the latent phase of a juvenile cancer, and the importance of discovering whether these changes are the same for cancers of the immune system (i.e. leukaemia and lymphoma which are known collectively as haemopoetic neoplasms) and other malignant diseases.

While still in Oxford George had heard me discussing the possibility that the postwar increase in leukaemia mortality might have been caused by modern drugs compensating for loss of immunological incompetence and thus preventing infection deaths whose underlying cause was an early effect of the cancer. He discovered that I had been assembling data on early deaths from pneumonia and leukaemia in three countries (US, Japan and UK) and asked to borrow these. A few weeks later he produced a statistical analysis of each data set which greatly strengthened my hypothesis since each one showed that by the time young children were on the brink of developing leukaemia they had become so infection sensitive that the risk of a pneumonia death is over 300 times greater than the normal risk. Therefore, here was both evidence of George's analytical skills and a warning that, when a serious infection coincides with the latent phase of a juvenile cancer, only children given antibiotics are at all likely to survive, and this survival will also depend upon whether this disease is a haemopoetic neoplasm or other type of cancer.

After the Birmingham move George had a short spell as consultant to the Mancuso study of Hanford workers. He used this opportunity to devise various ways of coming to grips with a problem which he was the first person to recognize. This was the problem of the same individual being exposed to a small dose of ionizing radiation not once but several times at constantly changing ages. George examined this problem from various angles and eventually decided that, although resistance to cancer induction effects of radiation is exceptionally high at twenty years of age, by fifty years it may be no greater than it was shortly before birth.

When Dr Mancuso lost his research contract (largely as a result of George's analysis of Hanford data producing risk estimates which were much higher than ones based on extrapolation of high dose effects) George resumed work on OSCC data and gradually assembled the evidence which underpins several important conclusions. For example, he showed that all victims of early cancer deaths are, from birth onwards, more infection sensitive than normal children, and that this undue sensitivity is most pronounced in cases of leukaemia. Therefore, when antibiotics are given to children who are (unknowingly) developing a malignant disease, an infection death may be prevented, but this 'narrow escape' will shortly be followed by a cancer death. This is probably why the increase in leukaemia mortality (which provided the original excuse for the Oxford survey) gathered momentum when penicillin began to replace sulphonamides. As was noted by David Hewitt at the time, this change affected affluent before poor societies and was all but confined to infection sensitive age groups. These groups include children under two years of age. However for these children low rates of leukaemia mortality were still being recorded which offered a striking contrast to the mounting death rate for three years olds. Hence the early peak of mortality which remains an unsolved problem.

THE OXFORD SURVEY OF CHILDHOOD CANCERS

The Oxford Survey is no longer alone in showing that (a) there is a cancer hazard associated with obstetric radiography and (b) the radiogenic cancers have a distinctive age distribution. But no other survey has tried to decipher the complex relationships which exist between cancers and infections, or to make a close study of inoculations in relation to the cancers.

The national campaigns which are responsible for the fact that most children in technologically advanced countries are now inoculated against several infectious diseases are not concerned with cancer prevention. Nevertheless as a result of innumerable children being given several boosts to the immune system, some early cancer deaths may have been prevented. Presumably this is the result of immune system effects halting the pathological processes set in motion by cancer inductions. Leastways, according to OSCC data any type of inoculation can prevent any type of cancer (though this is especially likely to happen when a relatively late inoculation coincides with a slow growing tumour). These observations suggest that there may be a crucial period between induction and diagnosis when any boost to the immune system is important. Therefore it is possible that premature infection deaths are not the only consequences of infections coinciding with cancer latent periods.

Finally, the reason why the spontaneous and radiogenic cases in the Oxford Survey proved to be so alike is gradually becoming clearer. This clarification is the result of comparing OSCC data with background radiation dose rates in various regions. In this way a cancer effect of the daily (and inevitable) exposures to background radiation was detected whose magnitude was compatible with this being the only common cause of early cancer deaths. This is too recent an observation for further elaboration but it shows that, even after thirty years, OSCC data still have their use. Therefore, let me end this account of achievements with a short postscript containing George's own description of his work and two ideas which may yet provide him with more work.

POSTSCRIPT

When asked by a colleague of Dr Mancuso to describe the work of our unit, George gave the following reply: 'It is my job to prove that Dr Stewart's theories are wrong.' Hence the strength of our long association.

One of the ideas which is still awaiting a 'George rebuttal' is that something akin to the cancer inhibitor effect of inoculations might be the reason why two unusual types of haemopoetic neoplasms (i.e. chloroma and Burkitt tumour) are found mainly in malaria infested regions. In these regions all children over two years of age have survived repeated attacks of malaria and related infections. Therefore, the whole population must consist of persons who, early in life, acquired immune responses which were so strong that they possibly placed constraints upon mutant cells as well as live pathogens. Whether such constraint is possible I do not know. But if it were we could account for chloromas and Burkitt tumours being unusually localized forms of haemopoetic neoplasms, and for there being exceptionally high levels of antibodies to certain viruses associated with Burkitt tumours.

My second idea is an attempt to account for the early peak of leukaemia mortality by suggesting that it might be a consequence of antibiotic intervention being (in relation to childhood infections) more successful in revealing the true prevalence of lymphatic than myeloid leukaemia. This requires the assumption that both myeloid and lymphatic cases with foetal origins prevent normal development of the immune system (which leads to premature infection deaths of children who would otherwise die from either type of leukaemia) and that the myeloid cases also prevent normal maturation of haemoglobin (which leads to early deaths from tissue anoxia of children who would otherwise die from myeloid but not lymphatic leukaemia). This sequence of events would require involvement of a common ancestral cell of erythrocytes and myelocytes in cases of myeloid leukaemia and would also require exceptionally high levels of foetal haemoglobin to have associations with three conditions, namely, very young cases of myeloid leukaemia, postnatal infections which are not responding to treatment and sudden infant deaths.

THE OXFORD SURVEY OF CHILDHOOD CANCERS

This attempt to explain the early peak of lymphatic leukaemia mortality brings my story full cycle. Therefore, it only remains for me to comment on the timeliness of your award. In Britain widespread use of ultrasound by obstetricians only began in 1975. Therefore to prove that these examinations do not have any cancer associations might have required data collection from all pre-1990 deaths. The present final year (for lack of funds) is 1992. However, your award will prevent such early closure and may help us to attract funds which are needed to complete this project. Therefore, on behalf of all previous helpers, including our case and control mothers, I again say thank you.

Science from the Third World: The Example of Endod in Overcoming Obstacles to a New Approach to Community Health

DR AKLILU LEMMA

December 1989

THIS YEAR MARKS the twenty-fifth anniversary since I first came across the potential values of endod, the Ethiopian soapberry plant, scientifically called *Phytolacca dodecandra*. This encounter was made in the town of Adwa, in the northern part of Ethiopia, during an ecological investigation of the distribution of particular freshwater snails that serve as the critical link in the transmission of schistosomiasis, a serious and rapidly spreading tropical disease that afflicts some 200-300 million people in Africa, parts of the Caribbean and South America, and Asia.

I observed that in areas downstream from where people were washing clothes with endod, there were more dead snails than anywhere else. Observing this phenomenon repeatedly, I collected some live snails from upstream and asked one of the women to put a bit of endod suds from her washbasin into the snail container. Shortly after, the snails shrank, passed a few bubbles of gas, bled and died. It was this original observation that led to the discovery and to many subsequent years of scientific study, hard work, frustration, hope, and at times, despair, which came to a climax with the honour of the 1989 Right Livelihood Award, for which I and all my colleagues are most grateful.

Back in our laboratory in Addis Ababa the first questions we eagerly wanted to answer were: What were the level and diversity of activities of the endod berries? Are the different species of snails that transmit schistosomiasis affected by it? What else does it kill?

We found that the sun-dried and crushed berries, when suspended and serially diluted in distilled water killed all the different species of snails in minute amounts (ten to twenty-five parts per million) within twenty-four hours. This was done using standardized test procedures so that others could easily repeat our work, as indeed they did, and our findings were shortly reconfirmed.

There are alternative molluscicides to endod, but they come at a high price. We do have safe and effective chemotherapeutic agents for the treatment of schistosomiasis, but people who have been treated in endemic areas become reinfected rapidly. Health education and latrine building could help, but this is a long-range prospect tied to a rise in the socio-economic standard of society as a whole. Therefore, the use of molluscicides to control the snails seems to be the major alternative.

Chemical compounds, such as copper sulphate, have been used as molluscicides for many decades in Egypt and Sudan as an effective means of controlling schistosomiasis. In the recent past, however, another more effective compound, the nicolsamid ethnolmine salt called Bayluscide, produced in West Germany, has been found to be more effective and is currently the only molluscicide recommended by the World Health Organization (WHO) for wide use. But this compound is very expensive—$25,000 to $30,000 per ton in hard currency. Partly as a result of this prohibitively high cost, most developing countries, especially those in Africa, are not doing anything about controlling this disease. In the meantime, the many well-intentioned agricultural development and water conservation schemes are serving as new reservoir sites for the snails to breed and the disease is spreading rapidly, but silently!

The implications of this phenomenon, as can be readily realized, are very serious. The role of endod here is to replace the expensive chemical molluscicides. Through the development and use of simple, appropriate, agronomic techniques and extraction and application procedures, people could easily grow, process locally, and use endod or similar other natural products to control schistosomiasis on a community self-help basis.

The first ten years, from 1964 to 1974, were the period of our greatest scientific challenges, leading to major discoveries and exciting prospects for multiple uses of endod. Most of the scientific studies on the molluscicidal and other properties of endod were undertaken during these exciting ten years. Findings were published in major international publications and presented at a number of international conferences.

The results of our early pioneering laboratory studies and field evaluations showed the very high potency and specificity of endod to snails and its safety to mammals and other non-target animals and the ecology. These studies were published in various journals, including the Bulletin of the World Health Organization in 1970 (vol. 42, pp. 597-612).

During a two-year sabbatical leave I spent in the United States (1970-1972), important chemical and toxicological investigations were undertaken. At the Stanford Research Institute (SRI) in Menlo Park, California, the following studies were conducted: chemical isolation and identification of the active principle, named 'lemmatoxin' by Robert Parkhurst, the chemist who did the work; tests that established the non-mutagenic nature of the product with additional studies on the safety of the butanol extract to non-target animals. Other potential uses of endod were also investigated: its fungicidal properties, its potential as an additive to detergent formulations, as a foaming agent in lightweight concrete preparations, and as a spermicide for possible use in birth control (for which SRI obtained a patent). I also spent some productive time at the University of California Medical Centre in San Francisco, where, in collaboration with Professor Donald Heyneman, we did additional toxicological and molluscicidal tests. At the Harvard University School of Public Health, working with Professor Andrew Spielman, we studied larvicidal effects of endod against mosquito larvae, as well as its snail ovicidal (egg-killing) properties. At the US Public Health Service in Puerto Rico, in collaboration with Dr Barnett Cline, we conducted a limited-scale field evaluation of the butanol extracts in natural bodies of water.

During the period of 1969 to 1974, a major five-year project was undertaken in the town of Adwa to control schistosomiasis, primarily by use of locally collected endod as a molluscicide, involving active community participation—a critically important factor. The results showed that transmission of the disease was significantly reduced from sixty-three to thirty-three per cent in the population of 17,000 people. Among 3,500 children from the ages of one to six years, the infection rate fell from fifty per cent to seven per cent (eighty-five per cent reduction). These results were presented at the International Conference on Schistosomiasis held in Cairo, Egypt, in 1975 and published in the proceedings of that conference.

Another milestone was achieved during 1976 to 1981, when the Netherlands government supported some important agrobotanical studies on endod. The studies were conducted by a Dutch scientist, Dr Charles Lugt, and his Ethiopian counterpart, Dr Legesse Wolde-Yohannes, my colleague and co-winner of this Award. Out of over five-hundred varieties of endod collected from all over Ethiopia, they cultivated sixty-five different strains and selected three varieties with especially high molluscicidal potency and high yield of berries. One of these strains, Endod-44, has been introduced and successfully cultivated in at least five African countries (Kenya, Tanzania, Zambia, Zimbabwe and Swaziland) and Brazil. It has also been chosen to serve as the standard 'reference' strain of endod for all toxicological and field evaluation.

The period between 1981 and 1986 was a transition period, a struggle to get international acceptance and increased support for our studies, often marked by frustration, delay, and disappointment. We planned and expected that this period would see the expansion of our earlier studies, increasing the agricultural production and promoting the large-scale application and field evaluation of endod as a molluscicide in integrated control of schistosomiasis on a community self-help basis.

For such activities the Ethiopian government needed foreign starter assistance with leadership from international health bodies like the WHO. Surprisingly, with complete disregard for our twenty years of research and the centuries of traditional use of endod as laundry soap, the WHO required further studies to confirm the safety of endod to humans and the environment before they would give clearance for its wide use. They insisted that the scientific work we did in Ethiopia and elsewhere must be repeated under standardized conditions following 'Good Laboratory Practices' (GLP) in internationally recognised institutions—meaning in developed countries. As a precondition for field evaluation, they required that this biodegradable natural product be subjected to the same rigorous toxicological tests required before any unknown synthetic chemical product can be registered as a pesticide. This, of course, meant the need for substantial financial resources to support such studies in countries with 'GLP'. But the WHO was neither providing any resources, nor even supporting our efforts to raise funds for the confirmatory studies which they required. I have no explanation for this bias.

The only reason given for not allowing field testing of endod was its toxicity to fish. It is a well known fact, however, that the same fish-killing property is shared by other molluscicides, including the only molluscicide approved by WHO for global use, the petrochemical byproduct Bayluscide. But the important point here is that edible fish and disease transmitting snails do not normally live in the same habitat. Edible fish normally reside in larger bodies of water, whereas the snails require slow flowing streams and canals or shallow shores of larger water bodies. As molluscicides are applied to these bodies of water, the fish become irritated and swim rapidly into safer waters. The snails, without such mobility, are the principal targets of the molluscicides. Small fish in shallow waters may be killed by molluscicides, but the fish repopulate from untreated bodies of water upstream, as well as from the egg-masses of the fish, which are not affected by endod. And, as was pointed out, since endod is biodegradable, its active principle rapidly decomposes, breaking down into inert and non-harmful material within a few days.

From a traditional acceptability point of view, endod is a typical example of a natural product that has been selected by society through centuries of its safe use for washing clothes. People in the Ethiopian highlands have in the past, continuing into the present, adopted and cultivated the endod plant (synonymous with soap) near their homes, using the berries as a laundering agent for the glistening white cotton shawls, the shama, which are characteristic of our culture. The fish-killing property of endod is also well known and traditionally accepted. In fact, people in rural communities use it as an intoxicant to collect edible fish.

PRESENT STATUS

The last few years have seen important breakthroughs that have countered many of the biases and obstacles raised against endod that for so long have gone through a vicious circle: no funds were made available for the rigorous toxicological tests required before endod could be officially field tested—and no field tests could be sanctioned until these tests were completed.

As part of the continued fight to gain recognition for the study of endod, in 1983 I organized the First International Workshop on Endod, held in Lusaka, Zambia. A large number of concerned scientists and interested Africans attended, with the meeting formally opened by the Prime Minister. It was both a scientifically and personally rewarding meeting for my colleagues and me, for those who have been engaged in endod studies for many years. It gave us strong moral support, encouragement and a real sense of collegial solidarity. The proceedings of that meeting, published in 1984 in book form by Tycooly International Ltd., continue to serve as a consolidated source of useful information on endod.

A Second International Workshop was held in Swaziland in 1986, the proceedings of which were also published in book form with the financial support of UNICEF. By the time of the Swaziland meeting, the Ethiopian strain of endod, E-44, had been introduced and experimentally studied in a number of countries including Swaziland, Zambia, Zimbabwe, Kenya, and Brazil.

The Third International Workshop is now planned and will take place in Ethiopia in May 1990.

Following the workshops in Zambia and Swaziland, the United Nations Development Programme (UNDP)/UN Fund for Science and Technology for Development, in co-operation with the International Development Research Centre (IDRC) in Canada, convened an Endod Toxicology Expert Group meeting in New York in February 1986. The objective of that meeting was to identify and remove the main obstacles against the trials and eventual application of endod. A second meeting of the same group was held at the WHO in Geneva in December of the same year. This Expert Group developed a standard procedure for preparing water extracts of endod for testing in different laboratories and delineated selective basic toxicological tests to be performed under 'GLP', as required by WHO.

IDRC has since commissioned some of the acute mammalian and eco-toxicity studies that are now being performed in Canada, basically repeating what we did several years before, but this time under 'Good Laboratory Practices'.

In a parallel, independent initiative, the Finnish International Development Assistance Agency (FINNIDA) is presently considering whether to fund the remaining toxicological studies required for the international registration of endod as a pesticide, following WHO procedures used for registering any chemical product. A co-ordinating committee has been established to oversee this process. FINNIDA is also providing support for agrobotanical and community-based studies in Zambia on a bilateral assistance basis.

In other developments, scientists in ten African countries and Brazil have established an Endod Technical Co-operation Among Developing Countries (TCDC) Network and are planning to have a newsletter to exchange their research findings and co-ordinate their work. In the Netherlands, a group of concerned and interested scientists have formed a 'Support Endod Group' that is doing extraction and toxicological evaluation work. In Florence, Italy, an Italian 'Technical Working Group on Endod' has put together a major project for collaborative agrobotanical studies in Ethiopia and Zimbabwe. They are now trying to obtain some funds for this important effort. Most of these initiatives require basic financial support.

LESSONS FOR THE FUTURE

Thinking beyond endod, some useful lessons can be drawn from these twenty-five years of struggle to help promote the future of science and technology in Africa.

We have learned the hard way that the root problems of scientific research in Africa are not only the lack of adequate facilities and funds, but also the biases of individuals and organizations in industrialized countries who occupy key positions to advise and influence the decisions of donor groups. Even our own government officials who rely on foreign assistance and the scientific support of such international organizations may be unduly pressured.

Sadly, there are still some biases and doubts in some Western establishments, that such a positive experience could come out of Africa. Except for occasional lip service, little credit is given to the wisdom of traditional societies in their ability to select, over long periods of time, such natural products as endod for their continual and demonstrably safe use. To modify and redirect this traditionally accepted product for the control of schistosomiasis (whose spread owes much to well-intentioned but epidemiologically disastrous project designs) is a major challenge.

Thorough toxicological evaluation should indeed be required of any product before it is used in field conditions. But such evaluation should also include the traditional knowledge of the people about such products. The logic of requiring the same toxicological tests from both a known biodegradable natural product and an unknown synthetic chemical product should also be questioned. In any case, such toxicological tests at least on natural products, should be undertaken in collaboration between industrialized and Third World countries, with the objective of helping the latter build its own capacity for such studies.

After three decades of post-colonial struggle, hopes and disappointments, I believe that the best future course for Africa is to invest in efforts to build on the endogenous capabilities of its own people. This can be done through raising self-esteem and self-confidence by creating a sense of respect for the wisdom and experience of its own traditional societies, perhaps through the integrated application of modern and traditional technologies. Future actions should also recognize and strengthen existing African scientists and research and training institutions. Africa's scientific and technological manpower must be increased by many fold. It is only through this type of major effort that we can promote sustainable and self-reliant development in Africa.

International collaborative and development aid should be more geared towards helping us help ourselves, however slow and frustrating the process may appear, rather than creating increased dependence, as has been the case so often in the past. Put very simply, development assistance should focus more on teaching us how to fish rather than giving us fish when we are hungry and then only when our cries are heard throughout the world through the mass media.

In conclusion, I would like to thank the so many friends and organizations in both developed and Third World countries who have been our genuine partners in this venture. I would particularly like to recognise the support and encouragement of Addis Ababa University and the Ethiopian government, various donor groups in the United States, the Netherlands government, IDRC of Canada, and the United Nations, UNFSSTO and UNICEF. I would also like to thank my wife, who for twenty-five years has given her unfailing support in my struggles for the international recognition of endod.

Finally, I want to thank the Right Livelihood Foundation for giving us this opportunity to speak our minds frankly and openly to an audience that can make a difference! May this occasion serve as a major encouragement to the many able Third World scientists, in whose name we would like to accept this Award.

Address by Dr Aklilu Lemma's Co-recipient
DR LEGESSE WOLDE-YOHANNES
December 1989

I AM VERY HONOURED to be co-recipient along with Dr Aklilu Lemma of the 1989 Right Livelihood Award. The endod story is truly one with a holistic approach to preventing a disease (schistosomiasis or bilharzia) that affects millions of poor people in developing countries. Such an award not only gives us a personal honour, but provides a tremendous boost and encouragement to our struggle in promoting the wide and safe use of endod as a poor man's remedy for a poor man's disease. Very often it is overlooked that the solution to the poor man's problem is close at hand to the poor man.

After the initial enthusiasm for endod, outside interest essentially disappeared. In the country-wide survey to find endod plants we had by 1982 collected sixty-five different eco-types. From this number we were able to identify three with high molluscicidal potency.

Agrobotanical and ecological adaption studies were started in 1982 to determine the best locations, soils and climates, as well as food crops that can be intercropped with endod shrubs.

After the two international scientific workshops on endod, held in Lusaka, Zambia in 1983 and Mbabane, Swaziland in 1986, collaborative efforts were encouraged more than ever before, particularly among African countries, in developing the endod plant in the spirit of Technical Collaboration among Developing Countries (TCDC). To facilitate this work the Institute of Pathobiology at Addis Ababa University was designated to co-ordinate the study, as well as to provide Endod-S for the toxicological studies and clonal material from Endod-44 growing in isogenic plots at our institute.

One problem until very recently was how to satisfy the issue of the lack of acceptable toxicological tests that are necessary for the registration of any new pesticide. The International Development Research Centre in Canada has provided funds to chemists at the Stanford Research Institute in California, Carleton University of

Canada and the University of Lausanne in Switzerland to identify an extraction protocol. They have also provided funds for the basic toxicology tests that are now being carried out in Canada and Ethiopia. The irony here is that we are dealing with a product which people have used to wash themselves and their clothing in Ethiopia for centuries, and if there was a problem it should have manifested itself in some way by now!

Providing there are no problems regarding Endod-S toxicology, I expect that in early 1990, field trials will begin in Ethiopia, Zambia and Swaziland.

The presentation of this Award will allow us to achieve our final goal more rapidly. That is, the cheap production and simple extraction of a safe molluscicide that can be used by the needy in the prevention of a debilitating disease.

In conclusion, I would repeat my thanks and deep appreciation for the Award, and know that all my collaborating colleagues would wish to join me in thanking the Foundation.

Building Communities: People's Housing
JOHN F. C. TURNER
December 1988

THE EYE OF THE BEHOLDER

SOME YEARS AGO, when almost all politicians and professionals regarded urban settlements built by low income people as 'slums', 'eyesores', 'cancers' and so on, two Englishmen were standing on a hillside overlooking a huge 'barriado', or self-organized and rapidly developing squatter settlement, on the outskirts of Lima, Peru. One Englishman was myself, an architect in the process of being de-schooled and re-educated by the experience of working with these city builders. The other was a visiting Minister of the British Government, who had asked for a guided tour to see for himself what he had heard about such settlements. Working with the 'barriada' builders had already taught me much of what I know about housing and local development, and they had affirmed my faith in the immense capabilities of people, however poor they may be. I naïvely expected the visiting Minister to be as encouraged as I was by the sight of so many people doing so much with so little. But the Minister was appalled. He viewed it as a monstrous slum threatening civilization itself, while I saw a vast building site and a developing city. We returned to the embassy in mutually bewildered silence. Only some time afterwards did I realize that what we see depends on where we stand. One person's problem is another person's solution.

Today, many government members and officers would share my view, probably most from Third World countries. But even now, twenty-five years later, after that episode and the first publications suggesting that the so-called 'problems' are also solutions, there are many who still feel disgust and pity for the poor rather than respect for how they survive, and admiration for what they do despite truly appalling hardship. Many governments still fear people's own organizations and actions, on which real development and the world's future depend. The overt suppression of local initiative by openly repressive régimes, and its inhibition by covertly repressive government, are to be expected from those who equate development with the centralization of power and wealth. Those who are genuinely concerned about global and local inequities could be truly caring by supporting people's hopes and their own action—by far the greatest resource for overcoming despair and grinding poverty.

However, most donations from the general public are still given mainly in response to images of despair. Using patronizing titles such as 'Give Me Shelter', the media still tends to present the poor as objects of pity, clutching begging bowls, and in helpless, dependent poverty. These insults to the poor probably do extract larger gifts from the conscience-stricken but uninformed populations of rich countries. Until awareness of both sides of the reality is more widespread, fund-raising for development and emergencies may still depend on such appeals. Front-line workers and observers in the field are naturally first to learn and to understand the realities. Increasingly, the mass media communicate what many voluntary aid organizations already know and what bi- and multi-lateral agencies are fast learning: that relieving sudden emergencies and the ongoing disaster of poverty depend on complementing, instead of ignoring the victims' own resources and priorities.

PEOPLE DO MORE WITH LESS

The experience and achievements recorded by the Habitat International Coalition's NGO Habitat Project (HIC) (part of the United Nations International Year of Shelter for the Homeless, 1987) show that the continuing disasters of generalized poverty and vulnerability can be mitigated and eventually eliminated. The HIC project carried out an international survey of initiatives for home and neighbourhood improvement. Over 300 such ventures were identified worldwide, 200 of which were from the Third World, and of these 200, twenty were selected for intensive study by the project. These case studies in turn formed the material for the book *Building the Community: A Third World Casebook* (1988). The studies point to ways and means by which the more dramatic forms of homelessness following earthquakes and floods, famines and wars can be drastically reduced even if they can never be eliminated altogether. By concentrating attention on the human resources of the poor, rather than on their often appalling conditions, the case reports highlight the necessity of supporting locally self-managed action. Only through government policies which enable people can the immense potential for development by people be realized. Knowledge of what even very poor people are capable of doing and of what even very rich states fail to do for those who cannot afford market prices, undermines the false claims of those who would have us believe that either the state or the market can substitute for those community-based initiatives of the people. Those who sacrifice people on the altars of the marketplace or the state can no longer claim that happiness tomorrow depends on frustration today. In fact, policies that inhibit personal and local initiative abort community-building on which our future depends.

The poor build for themselves an enormously greater number of homes and neighbourhoods than can ever be provided by public welfare and private charities. Between half and three-quarters of all urban settlement and homebuilding in the rapidly growing cities of the Third World are built by and for the poor themselves. This majority, usually four-fifths of the population, have no access to new housing supplied by commercial developers or public agencies. Donors to housing charities vastly outnumber those who are eventually sheltered by their gifts.

Most of the well-off, in rich and poor countries alike, must still be confronted with these simple facts. And these facts must be seen against the one, overwhelming fact: that all life on our already badly-damaged planet depends on all of us doing far more with much less. When we see great numbers of low-income people building and improving their communities, and at costs three or five times lower than those built for them, we must admit that we have a great deal to learn from those builders and from their enablers.

Current policies usually frustrate and disable people. An Argentinian squatter-builder once said, and as millions more, squatters or not, know only too well: 'There is nothing worse than being prevented from doing what one is able to do for oneself.' Enablement is the key. Neither bureaucratic mass housing nor the uncontrolled market can build communities and eliminate homelessness. But people can, when they have access to essential resources and when they are free to use their own capabilities in locally appropriate ways.

The word 'people' means everyone: infants and children, youth, the aged, and women, as well as men. The achievements described in the HIC Habitat Project are mainly community initiatives, not the products of any one age or sex group. The cases show that specific needs according to gender, age, health, ethnicity or culture are far more likely to be served through community-based programmes than through commercial developments or through government schemes in which people have no significant part. The cases also show how home and neighbourhood building depend more on women than on men, who always dominate the paternalistic forms of market- and state-based housing provisions. Centrally-managed organizations have more to gain from co-operation than from the competitive pursuit of empire-building.

ONLY PEOPLE CAN BUILD COMMUNITY

By approaching housing as an activity, a process involving everyone, along with most of the resources on which life depends, we have a paradigm for the world as a whole. This may seem far-fetched to those whose views are limited to the political and economic rivalry of market- and state-based systems. But the perspective through which these Third World initiatives are viewed reveals that there are three interlocking and interdependent systems, not two. The cases show us clearly that the answer to the housing question is no longer simply a choice between, or some combination of, two methods of provision—market and state. The conventional view of politics simply as conflict and compromise between free markets and central governments is a gross oversimplification of reality, leading to incomprehensible explanations. It is impossible to paint realistic pictures of the world as we see it with two primary colours, but with all three, it is relatively easy.

A true perspective shows all three dimensions. The new politics are about new relationships among the three interdependent systems: state, market, and community-based systems which are non-governmental and non-commercial. The perspective and principles that one can see more clearly in the harsher realities of the Third World are the same as those now widely reported and discussed in food production and nutrition, medicine and health maintenance, education and other spheres of vital human activity. Many terms coming into current use already refer to the emerging third system, or to aspects of it: 'civic society', 'the voluntary sector', 'the informal sector', 'la société civile' and 'el sector popular' among others. But it may not be a coincidence that our political vocabulary has no widely recognized term for the 'third sector' or system and that pyramidally-organized societies inhibit and largely ignore the role of women—the natural leaders of the vital third system. A new balance of powers at all levels and in all basic social activities is vital, for a workable and sustainable future.

As in any other view of real experience, the cases confirm the generally overlooked fact that most human, material and even financial resources are invested in homes and neighbourhoods. Dwelling environments occupy the greater part of all built-up areas. Most lifetime is spent in the home and neighbourhood. More energy is used for servicing, maintaining and building homes and local facilities than for everything else together. Collectively, of course, we spend more money on and in the home than everywhere else combined. So how we build and live locally is inseparable from the issues of human, economic and environmental degradation and development. 'Housing', conceived as a sector, like a slice of cake, is a dangerous abstraction. It is part of the mystifying jargon so effectively used by those who can profit from it, as long as the third system fails to express its autonomy and allows the state and market to take over.

When housing is usurped by commercial and political interests and powers, quantities are all that seem to matter. The qualities of housing, what it does for people, as distinct from what it is as a commercial or political commodity, have to take second place and are often ignored altogether. This is not due to corporate or bureaucratic perversity but to the fact that no large, centrally managed organization can possibly cope with the extreme complexity and variability of personal and local housing needs and priorities—demands that must be met if the housed are to invest their own time and effort in the acquisition, improvement and maintenance of their dwellings and surroundings.

HOUSING ECONOMY DEPENDS ON LOCAL AUTONOMY

If the satisfaction of a society's housing needs depends on the economic use of available resources, then it depends on people's own personal and local knowledge. As politicians are fond of saying, people are society's principal resource. But as politicians are less inclined to proclaim, the use of that resource depends on enabling policies that free and encourage people to use what they know and to do what they can. Individual and collective satisfaction depends on the release of personal and local knowledge, skills and initiative.

Knowledge depends on one's experience and, as I said earlier, on what one can see from where one stands. What an insider sees, looking outwards and up from a personal and local situation, is quite different from what an outsider sees, looking down from the expert's professional altitude. While the connections between one small place and its surroundings are clearly seen from above, they are not visible from within. Conversely, the vital details are difficult to see or too numerous to cope with when seen from above. When outside experts are responsible for making detailed housing decisions for centrally administered multi-family developments, they are bound to generalize, however much they may have studied their 'target populations'. The managements are also bound to limit the variations, in order to minimize their costs. These are the so-called 'economies of scale', which are really dis-economies, resulting from an appropriate scale for the job; and when people make their own personal and local decisions without due regard for the larger environment, substantial losses may also occur, for them and their neighbours.

These complementary kinds of essential expertise must work in co-operation in order to achieve an economic, convivial and environmentally sound use of non-polluting, renewable or long-lasting material resources. The relationship between the insiders and outsiders is critical. As their influence and effective authority over resource use is complementary and equal in practice, there must be sufficient equality to ensure mutual respect and the negotiation of any difference of opinion.

THE NECESSITY OF ENABLING POLICIES

The fact that so many people have done so much with so little in low-income countries while so little is done for low-income people with so much by their governments demonstrates the necessity of the radical policy changes which are taking place. Increasingly and with some perhaps vital assistance and encouragement from the international agencies and NGOs, Third World government policies are changing over from vain attempts to supply public housing to the support of locally self-managed initiatives. The necessity of enabling policies is not so obvious in countries whose governments can afford to subsidize all who cannot pay current market prices. But as the longer-term social and economic costs of depriving people of their freedom of choice and responsibilities undermine people's demands to be housed, we become increasingly interested in Third World experience and what it can teach.

In his address to the United Nations Commission on Human Settlements in Istanbul on May 5 1986, Dr Arcot Ramachandran, Executive Director of the UN Centre for Human Settlements, declared that: 'Our agenda for the next ten years must be to find the necessary capacities to apply [these] enabling strategies: [while we cannot be sure of success] we can only give a guarantee of failure for any other kind of strategy.'

Internationally, there is a growing acceptance of the fact that market-based, state-based and mixed housing supply policies have failed. The only alternatives are those based on the third sector or system which can be supported and enabled instead of being suppressed and disabled by market and state monopolies.

In the necessarily general and question-begging terms that one has to use in a summary, an enabling policy is a new balance between the complementary powers of the three systems—even where the third, people- and community-based system is badly eroded and weak as in Britain and most other highly industrialized and institutionalized countries. Dr Ramachandran's agenda implies a recognition of local capacities for deciding what to do locally, and of central capacities for enabling local self-management, by ensuring access to resources and for setting the limits to what may be done by people and their own community-based organizations and enterprises. Partnerships between these kinds of authority involve negotiation. The existence of mediating structures is therefore a pre-requisite for an enabling policy.

Non-governmental and non-commercial organizations—NGOs— and the community-based organizations, or CBOs, which they serve are essential. Only they can build up the necessary political pressures and only they can successfully balance opposing interests. Individuals and small groups are generally dependent on mediating organizations for successful negotiation. Ideally these are their own community-based organizations; more often people and their own CBOs depend on third party NGOs to assist in two vital ways:

▷ to help people organize, to articulate their demands, to assess their own resources, to plan and implement their own programmes and to manage and maintain their own homes and neighbourhoods;
▷ to act as mediators between people and their CBOs in their negotiations with the commercial enterprises and government agencies.

Only very small minorities can depend on NGOs to provide them homes or to improve their neighbourhoods—even smaller numbers than those who can expect government to do the same. In other words, NGOs can and do make essential contributions to changes of policy through the demonstrations of alternative ways and means of home and neighbourhood building—ways and means that show what industry and government can and must do in order to build a just and sustainable society.

The successful introduction of enabling strategies, the substitution of 'support policies' for conventional 'supply policies', confronts society with three major tasks:

▷ Articulating and disseminating demands for access to resources and freedom to use them in locally appropriate ways;
▷ Identifying, developing and disseminating the ways and means of stimulating and responding to demands for enablement; and
▷ Deepening and disseminating understanding of the new paradigm of development which underlies and is generated by these changes of demand and response.

These tasks involve everyone needing shelter, home and neighbourhood. This agenda is therefore addressed to everyone able and willing to act: to everyone with a mind and spirit of their own, in all cultures and of all ages.

The Seikatsu Club: Women and Co-operative Community in Japan

MACHIKO YAJIMA and NOBUHIKO ORITO

December 1989

BACKGROUND — AWAKENING WOMEN

JAPANESE WOMEN HAVE BEEN TRADITIONALLY ADMIRED for the virtue of obedience to the absolute authority of patriarchy. Nevertheless, since World War II, women have so changed that the joke was popular: 'The two things which have become stronger are women and nylon stockings.' This metamorphosis of women has two phases; before the 1960s, and after that time. In former times, the patriarchal system had collapsed through a series of democratic policies. Japanese women became free from the traditional family system. But although they were free legally, they were not yet free in the sense of social and economic conditions.

The industrialization of the 1960s resulted in a stronger position for the housewife in her household. Before I continue with an account of our Seikatsu Club Consumers' Co-operative (SCCC), I would like to explain the ordinary family life of the Japanese salary worker. Industrialization and economic growth spurred the tendency towards business centralized in large cities, and caused an increase in nuclear families living in urban areas. In their families the wives, who were already free from traditional patriarchy, became the proper manager of the household.

Meanwhile, the husband devoted himself to business in his company for the main purpose of earning a salary to sustain his family. It became the particular custom in the Japanese salary worker's family that the husband gave his whole salary to his household. It was the Japanese philosophy of life to endure the sufferings caused by severe inflation and the lack of family funds after the war. Under such conditions, the housewife naturally became the master of her family finances. Furthermore, the custom has been continued by the system where the company directly pays the husband's salary to a bank account which is open also to the wife. The modern banking system thus seems to support the wife's strong position in her household. Owing to the popularity of electrical appliances, housewives, moreover, have been able to enjoy more free time in which to gather social information. Yet their husbands must work like horses for their salaries, being separated from family and social life.

This was the situation in the 1960s, when the Minamata disease occurred in Kyushu. Local people, after eating fish contaminated with organic mercury from a local chemical industry, suffered terribly with pains. People, above all housewives, who were responsible for their family's well-being, rose up against this industrial pollution. They began to organize their own consumers' co-operatives to acquire safe products. The SCCC was one of them, and it has been developing alternative action against overindustrialized society.

THREE MAIN FIELDS OF ACTIVITY

Most of the 170,000 members of SCCC are ordinary housewives who take care of daily household matters. Through the co-operative movement of SCCC, we have become confident that we housewives can play a more active role in society. The Seikatsu Club is a consumer's co-operative association that started in 1968 when industrialization had been rapidly progressing in Japan. The aim of our co-operative movement was the development of an alternative lifestyle in contrast to mass production and mass consumption. We believe we will be able to realize this goal through the following three methods, showing:

1. Respect for life, through conservation of precious resources and protection of the environment by a collective purchasing movement;
2. Organization of women's 'shadow' work in their households through worker's collectives;

3. Development of a women's political movement based on citizens' voices in which participation in local politics is possible.

I will now briefly explain these.

COLLECTIVE PURCHASING AS A SOCIAL MOVEMENT

We started the co-operative movement twenty years ago by purchasing 200 bottles of milk, and now we have come to provide 400 products of our own. An ordinary Japanese family consumes 800 to 1,200 household products including food and clothing, and the Seikatsu Club covers one-half to one-third of them through collective purchasing. This collective purchasing works as a counter culture against wastefulness in society in the following three ways.

First, big companies advertise their products in various designs, sizes and shapes to increase consumption, but we need only one type of any given product. Soy sauce is packaged in numerous sizes and shapes in our country, but we provide nothing other than one litre glass bottles of thick soy sauce. Through limiting variety, the Club is able to streamline production and distribution. This prevents waste of material and also cultivates creativity in daily life.

Second, we think quality and safety of food is extremely important. By collective buying, we are able to bypass the commercial market and buy directly from producers. This does more than merely eliminate the middleman's added distribution cost. It enhances co-operation and awareness by keeping consumers in touch with the production process. Also, this system ensures freshness, which means that preservatives and additives are not necessary. Farmers do not need to use pesticides or chemical fertilizers which are harmful to our health and the ecological system. It also enables them to maintain organic farming. Incidentally, SCCC has been supporting a group of the families who have suffered from the Minamata disease through collectively buying the sweet summer oranges produced by them. They rehabilitated themselves by organic farming, or less agricultural chemical farming of oranges, so as not to be a polluting source themselves.

Lastly, we found that consumers had been damaging the water quality of rivers and seas by using synthetic detergent, which is an oil chemical product. This is a serious problem to the Japanese who eat the traditional daily meal of rice supplemented by fish and seaweeds. Therefore we have rejected all kinds of synthetic detergents, and have been using natural soap only. As a result, eighty per cent of agricultural co-operatives in Yamagata prefecture, which mainly provide us with rice, have succeeded in replacing synthetic detergent with natural soap. In Shiga prefecture, conscientious people working for the preservation of the water quality of Lake Biwa, the largest lake in our country, are using our soap. Fishery co-ops are also eager to follow us.

In the near future, two new member-factories for manufacturing soap are scheduled to be built. This soap is made from the cooking oil residue from members' own households and from others in their neighbourhoods, a method which has a dual effect on the environment: we can keep water oil-free and detergent-free. We are sure that this type of movement will spread to other areas, such as paper recycling, development of safer energy and substitute ingredients in aerosol cans.

HOW IS A 'HAN' RUN?
The 'Han', the basic unit for collective purchasing, is a co-operative group composed of seven or eight housewives in any members' neighbourhood. The idea of 'Han' originates in the communal organization for mutual aid in the traditional Japanese rural village. Mutual aid was a feature of life in a busy farming season, and in communal work everywhere in rural areas. Therefore, a 'Han' might be best understood as an organization which was transferred from the working system in traditional rural society.

Members make advance orders through a 'Han' once a month, and distribute products that are delivered to the house of the particular member on duty. Payment is also made through a 'Han'. The Seikatsu Club doesn't have shops because we know that the transaction at the counter tends to devalue the meaning of being a co-operative member. With no advertising, this system keeps distribution costs low, enables the prices of our products to be reasonable, and provides for the growth and maintenance of our co-operative movement. We believe that our basic business should be run fundamentally by our own shares. Therefore, each member makes a monthly saving of 1,000 yen (£4.50) until the amount reaches 80,000 yen (about £360).

A 'Han'is not only the basic unit for collective purchasing but it is also the smallest voluntary society involving a variety of positive elements. We can establish good personal relationships and can support each other when in need of help. On the other hand, when some personnel or communication difficulties occur, we try to solve these problems open-mindedly by ourselves. It means that we develop lessons for autonomy in a democratic society. 'Where there is no trouble, there is no progress.' So, a 'Han' may be the main component of co-operative community of the future.

ORGANIZATION OF SHADOW WORK BY WORKERS' CO-OPERATIVES

One step further from collective purchasing, our members have developed workers' co-operatives. The main motivation of this movement is our contribution to local community, using our experiences for collective purchasing together with our abilities outside the home. We were introduced to the idea of workers' co-operatives by the agenda, 'Co-operatives in the Year 2000', which was presented by the late Dr A. F. Laidlaw at the International Co-operative Alliance Congress held in 1980 in Moscow. We also learned about the co-operative movement of the Mondragon group in the Basque area in Spain, and the popular movement in the UK, the United States, and other countries. Then we decided to establish self-employed ventures, fitted to a self-reliance movement of Japanese housewives. In the Japanese labour market, chances for women to be employed have become equal to men's. But this is limited only to women in their twenties. It is very difficult for housewives to get jobs which they are interested in. By establishing self-employed ventures, it is possible for us to schedule time for household matters and work outside the home. This is even more possible if our house is near the workshop. Also, by applying the workers' co-operative system, we can organise the kinds of jobs which the government and big merchandise companies cannot serve. We have been establishing various types of ventures, such as serving lunches for workers in a community, or for the aged. The workers' collectives offer care and services to mothers before and after childbirth or to sick persons and their families. Some collectives provide a day service for working mothers. This is based on a barter system of work using labour tickets.

Besides these examples, some of our members organize recycling shops for buying and selling used materials such as refrigerators, furniture, clothing, sporting goods and so on. In the soap factories described above, labour is provided from workers' co-operatives as well as from non-members. In addition, we have other kinds of self-employed enterprises; restaurants, markets for fresh fish and vegetables, a bakery, a cookery school, a company to provide box lunches and school meals, and so forth. The number of these ventures now total eighty, and they employ some 2,000 women.

Each member of a workers' co-operative has a share in proportion to the business scale. Roughly speaking, each person has a share of 50,000 yen (£240) to 300,000 yen (about £1,345) in her venture. In order to help the establishment of self-employed ventures, the SCCC sometimes leases out its own extra rooms or lands, because the properties of SCCC are recognized as the common property of the co-operative community.

In summary, the workers' co-operative movement has many elements which are valuable for our modern society as an organization of informal shadow work typically seen in the home.

VOICES OF THE CITIZENS IN LOCAL GOVERNMENT
Finally there should be some explanation about our women's political movement. Needless to say, the Club members could not stop water pollution by themselves, by using only natural soap. So the members of SCCC carried out a petition calling for the elimination of synthetic detergents and brought it to the attention of the local government. However, at that time, the regulation was denied. We

decided, therefore, to send our own representatives to local assemblies. In 1979, the first Seikatsu Club member was successful in getting elected as a representative in Tokyo's Nerima Ward. In order to make our grassroots democratic movement more powerful, we organized a political network as a local action group. We call it the 'Seikatsusha Network'. Our main concerns are as follows:

▷ Protection of the environment;
▷ Institution of a welfare system in the local community;
▷ Establishment of a peaceful world without conflicts.

Under these themes, we are active in campaigns against the use of food additives, against using ports for missile-loaded battleships, for non-nuclear energy, for the declaration of non-nuclear cities, and so on. Especially worth mentioning is the Zushi City Network which has been struggling against the destruction of the green forest of Ikego Hills. The Japanese government has also offered this site to US military forces for the housing of missiles. In this case, activity promoting peace and protection of the green environment is equally important. We believe that if grassroots movements like this happen more, we can greatly influence the Japanese government. We will continue our activities honestly and steadily in our daily life under the slogan, 'Think globally, act locally'. Now we have thirty-six seats in twenty-nine city and town councils. Our representatives also support civil activities through women workers' co-operatives in co-operation with local authorities. All of them are free from any political party affiliations except for joining in local networks for a co-operative political movement. Seikatsu Club's Network contentions are very close to those of the Green Party in West Germany.

CONCLUSION

In Japan's continuing economic growth, the Seikatsu Club still plays a small part as an alternative influence. But we are not afraid to say what is true. Our slogan is 'Let's change our lives'. This means also to reform the situation as a result of joining our three fields together mentioned above—collective purchasing based on 'Han', services for the community by workers' co-operatives, and political participation through public pressure. We hope that the movement to build the co-operative community will diffuse throughout our country in the near future, and we are sure that this award will be an inspiration for improving the lives of not only Japanese but everyone on the earth.

Appendix

The Projects & Award Winners

1980–1985

THE PROJECTS & AWARD WINNERS
1980—1985

Hassan Fathy
1980
*'for saving and adapting traditional knowledge and practices
in building and construction for and with the poor'*

Dr Hassan Fathy is an Egyptian architect who has specialized in the promotion, teaching and upgrading of traditional building skills and the use of traditional materials.

He was head of the architecture section of the Cairo Faculty of Fine Arts and in 1969 was awarded the Egyptian Government's National Prize for Arts and Letters. He came to international notice with the publication in 1973 of *Architecture for the Poor* (University of Chicago Press). This book describes in detail Fathy's plan for building the village of New Gourna from mud and bricks employing traditional Egyptian architectural designs. Fathy worked closely with the people to tailor his designs to their needs, teaching them how to work with the mud bricks, supervising the erection of the buildings and encouraging the revival of ancient crafts.

In 1981 Fathy established the Institute for Appropriate Technology in Cairo as an institution to develop and apply his approach.

Hassan Fathy
4 Darb el Labanna Citadel
Cairo, Egypt

Stephen Gaskin
PLENTY International USA
1980
*'for caring, sharing and acting with and on behalf of
those in need at home and abroad'*

PLENTY USA is an international, non-profit, non-sectarian, relief, development, environment, education and human rights agency and is a Non-Governmental Organization associated with the Office of Public Information at the United Nations.

APPENDIX

PLENTY was founded in 1974 in the realization that all people are members of the human family and that, if we wisely use and share the abundance of the earth, there is plenty for everyone. From February, 1976 until the end of 1980, PLENTY employed more than 100 American volunteers in projects with the Mayan people of Guatemala (working on primary health care, potable water systems, soya bean agriculture and food processing, and communications technology).

While working with the Mayans in Guatemala, PLENTY made the strengthening and preservation of indigenous cultures a priority. 'We learned to what an amazing degree we shared the values and visions of these precious cultures, and that, for us, development was no longer a one way trip where we, the privileged, provided help to the under-privileged. We saw that, in truth, it was a fair exchange where every participant had something valuable to give.'

In 1978 the PLENTY Ambulance Service was established in the South Bronx, New York, providing free emergency medical care and training to the embattled residents of that sprawling American ghetto. That same year a rural village development programme was begun in tiny Lesotho, a country landlocked by South Africa.

Early in the 1980s PLENTY founded a free health clinic for Central American refugees in Washington, DC and undertook small-scale agriculture projects in Jamaica, St. Lucia and Dominica in the Caribbean.

Today PLENTY is involved in marketing Mayan textiles in the US for weaving co-operatives in Guatemala through its One World Trading Company; with environmental protection and nuclear issues through its Natural Rights Center; with Native American economic development primarily at Pine Ridge Reservation in South Dakota and among the Carib Indians in the West Indies; and, most recently, with a programme which is employing Masters of Business Administration graduates of American universities as volunteers in short-term projects. Two related organizations, PLENTY Canada, founded in 1977 and PLENTY Espana, founded in 1987, are engaged in similar activities.

Stephen Gaskin
41 The Farm
Summertown TN 38438
USA

Mike Cooley

Lucas Aerospace Joint Shop Stewards Combine Committee
1981
*'for designing and promoting the theory and practice
of human-centred, socially-useful production'*

Mike Cooley worked for many years in the aerospace industry as a senior design engineer and was an active trade unionist. In the early 1970s he became one of the pioneers of the now-famous Lucas workers' Corporate Plan, whereby Lucas workers threatened with unemployment organized across factory and union boundaries to draw up their own plan for socially-useful production, detailing 150 products which they and Lucas could make, including kidney machines, heat pumps, a road-rail bus and airships. The Plan was published in 1976 and, while it was rejected by Lucas management, it had a great impact in labour movement and other circles both in the UK and abroad. It led to the establishment of the Centre for Alternative Industrial and Technological Systems at the North-East London Polytechnic and similar units elsewhere in the UK. It has also resulted in other Worker Plans in other industries. Moreover, most of the technical ideas in the Plan have proved viable and been produced, though often with less emphasis on their social usefulness than the Plan envisaged.

In 1980 Lucas sacked Cooley, alleging too much time spent on union business or pursuing alternative technology interests. He became Director of the Technology Division of the Greater London Enterprise Board (GLEB), which had been set up by the Greater London Council to combat unemployment in London, and organized the London Technology networks, which link community groups, universities and polytechnics in the development of ranges of ecologically desirable products and systems which can then be used to establish new small businesses and co-operatives.

From his GLEB base Cooley has spun off a number of independent but related projects and activities: the London Innovation Trust, a charity to support the development of prototype products particularly for the disabled and otherwise disadvantaged; the Technology Exchange, which has built up a product databank of socially relevant and ecologically desirable products and which has recently received an EEC grant to link with similar networks in Spain, Denmark, Ireland, Greece and Greater Brussels; Twin Trading, which stimulates fair and mutually supportive trade between industrial and non-industrial countries; the journal 'AI and Society' which profoundly questions the drive towards artificial intelligence systems and the gradual displacement of human beings; and the ESPIRIT project 1217 which is in the process of designing and building a Human-Centred Advanced Manufacturing system in which human skills are enhanced rather than diminished and subordinated to machines, and which has pioneered the concept of Human Centred Systems. Cooley helped to set up the International Institute for Advanced research in Human-Centred Systems, of which he is now president.

Cooley has always been concerned to disseminate his ideas to the public at large and has made two films 'Look No Hands!' and 'The Factory of the Future' (supported by an educational pack), an eight part radio series 'Ways of Knowing' for Radio Telefís Eireann, and lectured in Australia, Japan, most European countries and the USA. He is also currently a Visiting Professor at the Universities of Bremen and Manchester.

Mike Cooley
Thatcham Lodge
95 Sussex Place
Slough
Berks SL1 1NN
UK

Bill Mollison
Permaculture Institute
1981
*'for developing and promoting
the theory and practice of permaculture'*

Bill Mollison was born in 1928 and has been called the 'father of permaculture', an integrated system of design encompassing not only agriculture, horticulture, architecture and ecology, but also money management, land access strategies and legal systems for businesses and communities. The aim is to create systems that provide for their own needs, do not pollute and are sustainable. Conservation of soil, water and energy are central issues to permaculture, as are stability and diversity.

Mollison's two early books *Permaculture One: A Perennial Agriculture for Human Settlements* (with David Holmgren, Transworld Publishers, 1978) and *Permaculture Two: Practical Design for Town and Country in Permanent Agriculture* (Tagari Publications, 1979) have sold over 100,000 copies and been translated into four languages. His new book, *Permaculture: a Designer's Handbook*, has recently become available from the Permaculture Institute. In addition, Mollison has written various articles and reports on permaculture for governments, educational and voluntary organizations and the general public.

The main focus of the Permaculture Institute is education. Since its inception in 1978, its certificated design courses have attracted more than 2,000 people, most of whom are now active in the practice or teaching of permaculture around the world. Independent permaculture institutes have been established in several countries and the movement is linked by biennial international conferences and the International Permaculture Journal.

Bill Mollison
Permaculture Institute
PO Box 1, Tyalgum
New South Wales 2484
Australia

Patrick van Rensburg
Education with Production
1981
'for developing replicable educational models for the Third World majority'

Patrick van Rensburg resigned his post as South African Vice-Consul in the Belgian Congo in 1957, in protest against the apartheid policies of the South African government. He subsequently joined the Liberal Party of South Africa and, with that party's support, became closely involved on a private trip to Britain in 1959 with the campaign to boycott South African goods, the Boycott Movement, which preceded the Anti-Apartheid Movement. Returning to South Africa his passport was confiscated and he was forced to flee the country following the State of Emergency after the Sharpeville shootings.

After a brief spell in Britain he went back to Africa in 1962 to Bechuanaland, which on independence from Britain became Botswana and of which he became a citizen in 1973. In Botswana he founded the Swaneng Hill School, and, following its success, two other schools in association with the Botswana government; the Swaneng Consumers' Co-operative; and the Brigades Movement. His experience with the schools and Brigades through the 1970s led in 1980 to his establishment of the Foundation for Education and Production.

Van Rensburg's education approach was radically different from usual practice. The curricula had a strongly practical orientation, including agriculture, building, typing, for example, as well as giving more traditional school instruction. While his teaching standards were as high as anywhere, the schools were low cost because of the traditional, frugal living standards they embraced. And the cost was further lowered, in an effort to bring the schools within reach of ordinary Botswanans, by the Brigades, which were self-help organizations of the students which produced goods and services both for the schools and for sale to help finance their education.

Patrick van Rensburg
Foundation for Education and Production
PO Box 20906
Gaborone
Botswana

Erik Dammann
The Future in Our Hands
1982
*'for challenging the values and lifestyle
of the West to promote a more responsible attitude
towards the environment and the Third World'*

Erik Dammann was born in Norway in 1931 and began his professional life in design, advertising and public relations. Becoming disillusioned with the consumerism his work required him to promote, he went with his family to live for a year in Polynesia, where they shared the life and lifestyles of the villagers in a culture founded on co-operation and sharing. He returned to Norway with the realization that the West's focus on competition for personal gain had more to do with social structure than human nature; with a deep respect for other cultures; and with a sense of responsibility for the way in which they were being systematically destabilized or destroyed by the consumption-oriented Western lifestyle and world view. He also perceived an immense gap between the stated values of Western society, such as justice, freedom, responsibility and solidarity, and the actual impact of that society on people in other countries and on the Earth.

Working on these issues following his return to Norway, Dammann published in 1972 *The Future in Our Hands* (English edition, Pergamon Press), a book which touched an enormously responsive chord in people both in Norway and elsewhere. In 1974 Dammann devoted himself full-time to establishing a Future in Our Hands movement, which grew during the 1970s to have over 20,000 members and considerable political influence. The movement has individual groups, a Development Fund that has funded support for projects in twenty developing countries, an Alternative Bank which gives loans for alternative development projects in Norway, and an Information Centre to promote political, personal and social change towards a more just and conserving society, as described in his later book *Revolution in the Affluent Society* (English edition, Heretic Books). The Future in Our Hands movement has spread to ten countries in both the industrial and non-industrial worlds.

Erik Dammann
Future in Our Hands
Torggata 35
0183 Oslo 1
Norway

Anwar Fazal
Consumer Interpol
1982
'for fighting for the rights of consumers and helping them to do the same'

Anwar Fazal, born in 1942, first became associated with the consumer movement in 1968 and worked on consumer affairs for, among others, the Government of Mauritius, the Hong Kong Consumer Council and the United Nations Food and Agriculture Organization. In 1978 he was the first person from the Third World to become President of the International Organization of Consumers' Unions, (IOCU), an independent, non-profit group which links the activities of consumer organizations in over half the countries of the world. IOCU promotes international co-operation in consumer protection and education, represents the consumer interest at the global level, furthers the dissemination and documentation of consumer-related information and facilitates the comparative testing of consumer goods and services.

Over the next four years Fazal galvanized the international consumer movement, founding a number of global consumer networks which he calls 'a new wave of consumer movement'.

Anwar Fazal
International Organization of Consumers' Unions
PO Box 1045
Penang, Malaysia

Petra Kelly
1982
'for forging and implementing a new vision uniting ecological concerns with radical disarmament, social justice and human rights'

Petra Karin Kelly, born in 1947, was one of the founders in 1979 of Die Grünen, the West German Green Party, which she has described as 'a non-violent ecological and basic-democratic anti-war coalition of parliamentary and extra-parliamentary grassroots oriented forces within the Federal Republic of Germany'. As one of the party's national Chairpersons from 1980 to 1982 she achieved international renown as the German Greens put green politics on the European political agenda in the early 1980s. In 1983 she was elected to the German Parliament as one of 28 Green MPs, was speaker of the Green Parliamentary Group until 1984 and has been a member of the Foreign Relations Committee since 1983. She was re-elected to the Bundestag in 1987.

Kelly has concentrated in her political work and speechmaking on the four themes closest to her heart: peace and non violence (she is Speaker of the German Federation for Social Defence), ecology, feminism, and human rights, and the links between them. She believes in civil disobedience and has participated in many such actions in many parts of the world. In a different area, Kelly also founded and since 1973 has chaired the Grace P. Kelly Association for the Support of Cancer Research in Children, a Europe-wide citizen action group against childhood cancer funded after the death from cancer at the age of 10 of Petra's sister, Grace. Kelly's first book *Fighting for Hope* was published in 1984 (in English by Chatto and Windus). She has written subsequent books on Hiroshima, childhood cancer and Tibet.

Petra Kelly
Bundeshaus
5300 Bonn 1
West Germany

Participatory Institute for Development Alternatives
1982
*'for developing processes of self-reliant,
participatory development in the Third World'*

PIDA is a non-governmental development organization which was established in 1980 for the purpose of initiating and promoting grassroots participatory development processes in Sri Lanka. PIDA's approach to development grew out of the pioneering work of Asian scholars in the mid-seventies (recently republished as *Towards a Theory of Rural Development* by De Silva et al, Progressive Publishers, Lahore, Pakistan 1988) centred on the UN Asian Institute of Development in Bangkok. These scholars initiated a process of reflection on the reality of Asian poverty and the failure of past developmental efforts, and attempted to develop a conceptual framework for an alternative development in Asia.

To turn this theoretical work into practice, the Sri Lanka government and the United Nations Development Programme began a Change Agents Programme in 1978, which in turn led to PIDA's foundation.

PIDA
32 Gotami Lane
Colombo 8
Sri Lanka

APPENDIX

Sir George Trevelyan
Wrekin Trust
1982
*'for educating the adult spirit to a new
non-materialistic vision of human-nature'*

Sir George Trevelyan, Bart., has been characterized in a UK national newspaper as 'a spiritual leader of the New Age movement at what he describes as the most exciting and most critical period in man's history'.

Born in 1906 he was in fact an agnostic until he attended a lecture on Rudolph Steiner in 1942. Thereafter his study of anthroposophy profoundly altered his philosophy of life and laid the basis for much of his future work.

After the Second World War he went into adult education as Principal of Attingham Park, Shropshire, where he first started giving courses on the spiritual nature of man and the universe, and, on his retirement in 1971, founded the Wrekin Trust to continue this work. An educational charity, the Wrekin Trust is not affiliated to any particular doctrine or dogma, nor does it offer any one way to 'the truth'. Rather it helps people find the disciplines most suited to them, organizing conferences on the holistic world view and introductory approaches to various disciplines, as well as offering a curriculum for ongoing spiritual training. The inspiration is derived from the medieval concept of the University, which was concerned to find and orchestrate methods and systems of knowledge leading to union with the One, as the term 'Universe' (turned to the 'One') reveals.

Trevelyan has written three books: *A Vision of the Aquarian Age* (1977), *Operation Redemption* (1981), and *Summons to a High Crusade* (1985). In addition to help to plan and contribute to the work of the Wrekin Trust, Trevelyan continues to develop lecture tours in many parts of Britain and abroad.

Sir George Trevelyan
The Old Vicarage
Hawkesbury, Badminton
Avebury
Avon GL9 1BW
UK

Leopold Kohr
HONORARY AWARD
1983
'for his early inspiration of the movement for a human scale'

Leopold Kohr was born near Salzburg in Austria in 1901. He was educated at the universities of Innsbruck, Paris, Vienna and the London School of Economics. After a variety of occupations, including correspondent in the Spanish Civil War, he entered academic life, teaching first at Rutgers University in the US and then as Professor of Economics and Public Administration at the University of Puerto Rico from 1955-1973. He then taught political philosophy at the University College of Wales, Aberystwyth.

He was an originator, and for two and a half decades the solitary advocate, of the concept of the human scale and the idea of a return to life in small communities, both of which were later made so popular by his friend Fritz Schumacher in *Small is Beautiful*. He consistently advocated the effectiveness of the small autonomous unit in the solution of human problems. With regard to Third World nations, he early asserted that massive external aid crippled their vital communal identity and stifled local initiative and participation. Instead he advocated a dissolution of centralized structures in favour of a system of small communities solving local problems with their own material and intellectual resources.

These ideas were powerfully expressed in a series of books, including: *The Breakdown of Nations* (Routledge & Kegan Paul, 1957), *Development Without Aid* (1973) and *The Overdeveloped Nations* (1977), both published by Christopher Davies, Swansea.

Leopold Kohr
Furstenweg 29
A-5034 Salzburg-Hellbrunn
Austria
and
170 Reservoir Road
Gloucester GL4 9SB
UK

APPENDIX

Amory and Hunter Lovins
Rocky Mountain Institute
1983
'for pioneering soft energy paths for global security'

Hunter and Amory Lovins work together as analysts, lecturers and consultants on energy, resource and security policy in over 15 countries. Their prophetic analyses caused Newsweek to place them among 'the Western world's most influential energy thinkers' and their work is noted for original syntheses, meticulously documented calculations and an emphasis on hardheaded economics.

Hunter Lovins has degrees in Law, Political Studies and Sociology and is a member of the California Bar. For six years she was Assistant Director of the California Conservation Project. Amory Lovins is a consultant experimental physicist educated at Harvard and Oxford who has published a dozen books and over a hundred papers, held various academic chairs, served on the US Department of Energy's senior Advisory Board, worked with Commissions and utilities in over thirty US states and consulted for financial institutions, utilities, private industries, governments and international organizations. His best known book is probably *Soft Energy Paths: Toward a Durable Peace* (Harper Colophon, New York, 1979). Together the Lovinses have co-authored four books and consulted, lectured and advised across the US and in over fifteen countries abroad. In 1982 they founded Rocky Mountain Institute (RMI) of which they are currently President and Vice-President.

Hunter and Amory Lovins
Rocky Mountain Institute
Drawer 248
Old Snowmass, CO 81654
USA

Manfred Max-Neef (CEPAUR)
1983
'for revitalizing small and medium-sized communities, fostering self-confidence and reinforcing the roots of the people'

Manfred Max-Neef is a Chilean economist who taught economics at the University of California (Berkeley) in the early sixties and later served as a Visiting Professor at a number of American and Latin American universities. He has worked in projects in Latin America, as an expert in social development with the Pan American Union, as a general economist with FAO and as a Project Manager with ILO. He has also written extensively on development alternatives.

In 1981 he wrote the book for which he is best known, *From the Outside Looking In: Experiences in Barefoot Economics*, published by the Dag Hammarskjöld Foundation. Late in 1981 he set up his organization CEPAUR (Centre for Development Alternatives) in Chile, where he still works.

CEPAUR is largely dedicated to the reorientation of development in terms of stimulating local self-reliance and satisfying fundamental human needs and, more generally, to advocating a return to the human scale.

Manfred Max-Neef
CEPAUR
Casilla 27.001
Santiago 27
Chile

High Chief Ibedul Gibbons
1983
'for upholding the democratic constitutional right of Palau to be nuclear free'

Palau, also called Belau, is a Micronesian archipelago in the Western Pacific with a population of 15,000 people. After World War II, it became a US Trust Territory by which the US was mandated to provide for the development of the political, social and economic well-being of Palau leading to its independence.

In 1979 the people of Palau voted overwhelmingly for a constitution which prohibited the use, testing, storage or disposal of nuclear, chemical and biological weapons, and the entry of both nuclear-powered and nuclear-armed ships and aircraft. Since then the country has been in continuing crisis. Its people have been forced to vote ten times in referenda concerning Palau's political status. Two presidents have died by shooting, political activity has been characterized increasingly by violence, intimidation and corruption, and the Palauan and

US governments have been accused of tampering with the democratic process.

At the centre of the crisis is the proposed Compact of Free Association between the United States and Palau, a concept discussed and widely supported in Palau during the 1970s. The Compact would give Palau internal self-government, a fifty-year guaranteed aid package and certain defence guarantees. In return, Palau would forego some foreign policy autonomy and allow the US certain military rights. These rights were incompatible with the anti-nuclear clause in the constitution and the US has refused to agree to the Compact while this clause is in force. This has led to a split in Palau between supporters of the constitution who want the anti-nuclear provision retained, and opponents of the Compact who want it repealed.

Chief Ibedul Gibbons received the Right Livelihood Award in 1983 on behalf of the people of Palau for the courageous defence of Palau's constitution against the violent political machination of Compact supporters. Support for the constitution is now being co-ordinated by the Belau Pacific Centre (addresses below).

Ibedul Gibbons
Address for information:
Bernie Keldermans
PO Box 1405, Koror
Palau 96940, via USA
or
Michelle Syverson
1027 Swathmore Avenue
Suite 135
Pacific Palisades, CA 90272
USA

Iman Khalifeh
HONORARY AWARD
1984
'for inspiring and organizing the Beirut peace movement'

Iman Khalifeh was born in 1955 and was educated in Beirut both before and during the civil war that erupted in 1975 and which still continues. While a teacher at the Nursery School of Beirut University College, at which she was also a research assistant and teacher trainer, she had the idea early in 1984 of organizing a peace march at which the silent majority of Lebanese who were against the war could express their protest and revulsion against the nine years of death and destruction through which they had been forced to live.

The idea caught on. The poem written by Khalifeh suggesting the march was printed in most of Beirut's newspapers. There seemed to be an enormous upsurge of popular feeling for the march and against the war. Perhaps for this very reason there was also an upsurge in the fighting between the rival militias the day before the proposed march, which was called off, to prevent any more casualties. A petition which was circulated instead collected over seventy thousand signatures very rapidly.

> We want simply to live in peace
> We want to raise up our children
> And save our brothers and sisters . . .
> We want our families to remain whole
> Let us walk out of our isolation and join one another . . .
> Let us walk out of our tears and screams of pain
> And hold together our only
> Slogan: No to the war
>   No to the tenth year
>   Yes to life.

(from Iman Khalifeh's mobilizing poem)

Iman Khalifeh
c/o Beirut University College
PO Box 13-5053
Beirut, Lebanon

Winefreda Geonzon

Free Legal Assistance Volunteer Association

1984

'for protecting the human rights of poor prisoners and the victims of injustice in the Philippines'

Winefreda Geonzon is a Filipino lawyer who in 1978 became the Legal Aid Director of the Integrated Bar of the Philippines in Cebu City, which brought her into direct contact with the many injustices and abuses of the legal system which occurred during the martial rule years of the Marcos régime. People, including young children, were jailed without charges or trial; imprisoned beyond their term; tortured and brutalized; or simply forgotten in prison.

In response Geonzon set up the Free Legal Assistance Volunteers Association (Free LAVA) as a free legal aid office for victims of violations of human rights, poor prisoners who could not afford to hire a lawyer, and for people whose cases had implications for social justice. Free LAVA sought to serve the many prisoners in these categories actually by seeking them out and visiting them in jail where its volunteers had first hand experience of the appalling prison conditions to which inmates were subjected: chronic overcrowding (Cebu City's main jail, built for 250, had a growing population of over 700 in 1987); poor or no sanitary facilities; cement floors with no sleeping mats, and little protection from inter-inmate assault or violence.

> Free LAVA
> Room 207
> 2nd Floor Mingson Building
> Cor. Juan Luna & Zamora S
> Cebu City
> Philippines

Wangari Maathai
Green Belt Movement
1984
'for converting the Kenyan ecological debate into mass action for reforestation'

Professor Wangari Maathai was born in Kenya in 1940 and was trained in veterinary anatomy, receiving her doctorate in 1971. She became Chairman of Veterinary Anatomy and Associate Professor of Anatomy at Nairobi University in 1976 and 1977 respectively.

Maathai has also long been active in the National Council of Women of Kenya, of which she has been Chair since 1980, and it was in the National Council of Women that the idea of the Green Belt Movement, a broad-based, grassroots tree-planting activity, was born. Its first trees were planted on June 5th (World Environment Day), 1977.

The Green Belt Movement grew very fast. By the mid-1980s Maathai estimated that it had about 600 tree-nurseries, involving and earning income for 2,000–3,000 women; had planted about 2,000 green belts of at least 1,000 trees each, involving about half-a-million schoolchildren; and had assisted some 15,000 farmers to plant private green belts. Maathai is currently taking forward a proposal with the United Nations Environment Programme for a Pan-African Green Belt movement, to spread the successful Kenyan experience to twelve other African countries.

Wangari Maathai
Green Belt Movement
National Council of Women of Kenya
PO Box 43741
Ragati Road
Nairobi
Kenya

Theo van Boven

HONORARY AWARD
1985
'for speaking out on human rights abuse without fear or favour in the international community'

Theo van Boven was born in the Netherlands in 1934 and obtained a doctorate in law in 1967. He was for ten years, until 1977, a lecturer in human rights at the University of Amsterdam, and from 1970-75 was the Netherlands' representative on the United Nations Commission on Human Rights. From 1977-1982 he was Director of the UN Division of Human Rights. Since then he has been Professor of Law at the University of Limburg. In addition, van Boven has served on numerous councils and committees dealing with human rights, including the Council of the International Institute of Human Rights (France) and the European Human Rights Foundation (UK), of which he is the Chairman.

As Director of the UN Division of Human Rights van Boven argued consistently that concern for human rights should not be a marginal activity within the UN system, but should become the core element of development strategies at all levels. He was concerned also to identify the root causes of human rights violations in connection with the development process, patterns of economic and political domination, militarization of societies and racial discrimination. In addition, he worked hard to strengthen the links of his office with non-governmental organizations.

His uncompromising approach to these matters led to major policy differences with the UN Secretary-General which led to his UN contract being terminated in May 1982.

Theo van Boven
Kantoorweg 5
6218 NB Maastricht
The Netherlands

Janos Vargha
Duna Kor
1985
'for working under unusually difficult circumstances to preserve a vital part of Hungary's environment'

Duna Kor, meaning the Danube Circle, was set up in 1984 as an environmental movement opposing the construction of an enormous dam and hydroelectric complex on the Danube. Its founder was Janos Vargha, a 40 year-old biologist who had earlier worked for some years for the Hungarian Academy of Sciences and was then on the editorial staff of the Magyar edition of *Scientific American*.

The $3 billion-plus complex was to be jointly built by Hungary and Czechoslovakia, with a considerable financial input from Austria, and involved drastic interference with nearly 200 kilometres of river; the flooding of fifty islands and 120 square kilometres of forests and fields; and the loss of valuable wildlife habitats. It also had incalculable implications for the groundwater of the region and the drinking water supply of some three million people. It had no convincing economic rationale and was opposed for this reason by the Presidency of the Hungarian Academy of Sciences in 1983. But the Hungarian Government was determined to proceed.

Duna Kor was a social innovation as well as a protest movement. Such groups were officially very much discouraged at the time it was set up and there was no chance of registering the group formally. Moreover, for over a year in 1984-85 no one was permitted to publish anything on the power project.

Recently, of course, the political climate in Hungary has changed and in February 1989 Duna Kor finally achieved its official registration. Duna Kor's influence and environmental awareness generally has grown enormously since 1984, as evidenced by a new 130,000 signature petition for a referendum on the power plant, which it organized in October 1988. This is still being considered, but by spring 1989 had not been rejected, by the government. Duna Kor is pressing on with the organization of local groups and is taking up other environment issues, always working from a strong scientific base as an interface between science and social action. The Danube itself has in many people's eyes become a symbol for the Hungarian political, social and economic reform process. If the Danube can be saved, so too can Hungary.

Janos Vargha
Utca 4
H 2097 Pilis
Borosjeno
Hungary

Judit Vasarhelyi
Soros Foundation Committee
PO Box 34
Budapest H 1525
Hungary

Lokayan
1985
'for linking and strengthening local groups working to protect civil liberties, women's rights and the environment'

Lokayan, meaning 'Dialogue of the people', is a research and documentation initiative which provides a forum for non-party political activists, through dialogues, workshops, working groups and lectures. These activities involve a large network of concerned intellectuals, activists and opinion makers. Through these, the Lokayan programme has sought to evolve a systematic critique of the established models of development and the state and to promote political action towards a new ideological crystallization suited to the needs and aspiration of the people.

Drawing upon a large variety of 'micro' initiatives that are struggling to achieve a just society, Lokayan aims at building a body of knowledge and opinion and concrete strategies of intervention at the 'macro' level that will promote a decentralized democratic order, enhance respect for cultural and social diversity of marginalized sections of society, and empower them all for participation in the larger movement for social transformation. The thinking, values, aspirations and experiences of these people provide the basis for alternative cultural, economic and political thinking and contribute towards minimizing the fragmentation that divides the various movements for change.

Lokayan publishes regular bulletins of comment, analysis and book reviews, which express the wide range of its concerns.

Rajni Kothari
13 Alipur Road
Delhi 110054
India

Cary Fowler and Pat Mooney
1985
'for helping the Third World preserve its genetic plant resources'

Cary Fowler was born in 1950, studied at Simon Fraser University, Vancouver, and at Uppsala University in Sweden, and is the author and editor of numerous publications on the world food crisis and on the erosion of the world's genetic resources.

Pat Mooney, born in 1947, has spent most of his life on the Canadian prairies and in agricultural development work in Asia, Africa and Latin America. In the mid-1970s he became increasingly concerned about the loss of agricultural genetic resources and in 1979 he published a report on the subject, *Seeds of the Earth*. This was followed in 1983 by his study *The Law of the Seed: Another Development and Plant Genetic Resources*, which attracted wide international attention when it appeared as a special issue of 'Development Dialogue' (1983: 1-2), published by the Dag Hammarskjöld Foundation in Sweden.

Fowler and Mooney have worked together since 1975. As international advocates for genetic conservation they have initiated worldwide educational campaigns on the problems of increasing genetic uniformity in agriculture and have proposed far-reaching conservation programmes. One of their proposals was the establishment of international seed banks, a plan which was adopted by the UN in 1983.

THE PROJECTS AND AWARD WINNERS 1980–1985

Since 1978 Fowler and Mooney have worked with Rural Advancement Fund International (RAFI), a small, non-profit organization which focuses on the socioeconomic impact of new technologies on rural societies. RAFI both undertakes extensive research into the science and corporations powering new technologies, and emphasizes practical action at both the international level and through grassroots organizing with those who will be most affected by technological change. Thus RAFI played a major role in stimulating the creation of the FAO Commission and Undertaking on Plant Genetic Resources and the International Fund for Plant Genetic Resources. At the same time, RAFI has organized with Third World partners numerous workshops in Africa, Asia and Latin America addressing both the global political concerns and the need for local farmers to secure their own crop genetic diversity.

Cary Fowler
RAFI (USA)
PO Box 1029
Pittsboro
North Carolina 27312
USA

Pat Mooney
RRI No 1
Beresford
Brandon
Manitoba R7A 5Y1
Canada

Bibliography

Bertell, R. *No Immediate Danger: Prognosis for a Radioactive Earth* (Womens Press, London: 1985).

Cooley, M. *Architect or Bee?* (Chatto and Windus, London: 1987).

Dag Hammarskjöld Foundation *What Now? Another Development* (Uppsala, Sweden: 1975).

Dag Hammarskjöld Foundation *Another Development: Approaches and Strategies* (Uppsala, Sweden: 1977).

Dammann, E. *The Future in our Hands* (Pergamon Press, Oxford: 1972).

Dammann, E. *Revolution in the Affluent Society* (Heretic Press, London: 1984).

De Silva, G., Haque, W., Mehta, N., Rahman, A., Wignaraja, P. *Towards a Theory of Rural Development* (Progressive Publishers, Lahore, Pakistan: 1988).

Durr, H.-P. *Proposal for a World Peace Initiative*, endorsed by the Council of International Physicians for Prevention of Nuclear War, June 1986 (Köln, West Germany: 1986).

Ekins, P. *The Living Economy: a New Economics in the Making* (Routledge and Kegan Paul. London: 1986).

Ekins, P. *New Movements for Social and Economic Change* (M.Phil thesis, Department of Peace Studies, University of Bradford: 1990).

Ekins, P. 'Green Ideas on Economics and Security' in Friberg, M. (ed.) *New Social Movements in Western Europe* (Padrigu Papers, Gothenburg University, Gothenburg, Sweden: 1988).

Fathy, H. *Architecture for the Poor* (University of Chicago Press: 1973).

Fowler, C. and Mooney, P. 'The Laws of Life: Another Development and the new Biotechnologies' *Development Dialogue* 1-2 (Dag Hammarskjöld Foundation, Sweden: 1988).

Galtung, J. *Essays in Peace Research* 5 vols (Ejlers, Copenhagen: 1974-1980).

Jungk, R. *Brighter than a Thousand Suns* (Penguin, London: 1960).
[Further information about Robert Jungk's work can be obtained from International Futures Library, Imbergstrasse 2, A 5020 Salzburg, Austria.]

BIBLIOGRAPHY

Kelly, P. *Fighting for Hope* (Chatto and Windus, London: 1984).
Kohr, L. *Development without Aid* (Christopher Davies, Swansea: 1973).
Kohr, L. *The Overdeveloped Nations* (Christopher Davies, Swansea: 1977).
Moore-Lappé, F. *World Hunger: Twelve Myths* (Food First Books: 1986).
Lovins, A. 'Energy, People and Industrialisation', Paper commissioned for the Hoover Institution Conference 'Human Demography and Natural Resources', February 1989 (Stanford University, California: 1989).
Lovins, A. *Soft Energy Paths: Toward a Durable Peace* (Harper Colophon, New York: 1979).
Max-Neef, M. *From the Outside Looking In: Experiences in Barefoot Economics* (Dag Hammarskjöld Foundation, Sweden: 1981).
Max-Neef, M., Elizalde, A., Hopenhayn, M. 'Human Scale Development: an Option for the Future' in *Development Dialogue 1989:1* (Dag Hammarskjöld Foundation, Sweden: 1989).
Mollison, B. with Holmgren, D. *Permaculture One: A Perennial Agriculture for Human Settlements* (Transworld Publishers: 1978).
Mollison, B. *Permaculture Two: Practical Design for Town and Country in Permanent Agriculture* (Tagari Publications: 1979).
Rensburg, P. van *Report from Swaneng Hill: Education and Employment in an African Country* (Dag Hammarskjöld Foundation, Sweden: 1974).
Seikatsu Club *The Seikatsu Club: Autonomy in Life* (Seikatsu Club, Tokyo: 1988).
Stewart A. *Half a Century of Social Medicine: An Annotated Bibliography of the Work of Alice M. Stewart* (edited by C. Renate Barber, Billingshurst: 1987).
Survival International, 'Resistance to Dams Holds Promise' in *Survival International News* No. 25 (S.I., London: 1989).
Trevelyan, G. *Summons to a High Crusade* (Findhorn Press: 1985).
Turner, John F. (with Robert Fichter) *Freedom to Build: Dweller Control of the Housing Process* (Macmillan, London: 1972).
Turner, John F. *Housing by People: Towards Autonomy in Building Environments* (Marion Boyars, London: 1972).
Woodhouse, T. *People and Planet* (Green Books, Bideford: 1987).